Undergraduate Texts in Mathematics

Readings in Mathematics

Editors
S. Axler
F.W. Gehring
K.A. Ribet

Springer Science+Business Media, LLC

Graduate Texts in Mathematics
Readings in Mathematics

Undergraduate Texts in Mathematics
Readings in Mathematics

Reinhard Laubenbacher
David Pengelley

Mathematical Expeditions
Chronicles by the Explorers

With 94 Illustrations

 Springer Science+Business Media, LLC

Reinhard Laubenbacher
David Pengelley
Department of Mathematical Sciences
New Mexico State University
Las Cruces, NM 88003-0001
USA

FRONT COVER ILLUSTRATIONS: *Background*: Sophie Germain's 1819 letter to Carl F. Gauss about her work on Fermat's Last Theorem. *Foreground* (clockwise from top left): Georg Cantor, Pierre de Fermat, Evariste Galois, Gottfried Leibniz; at center, Nikolai Lobachevsky.

Mathematics Subject Classification (1991): 01-01, 04-01, 11-01, 12-01, 26-01, 51-01

Library of Congress Cataloging-in-Publication Data
Laubenbacher, Reinhard.
    Mathematical expeditions : chronicles by the explorers /
Reinhard Laubenbacher, David Pengelley.
        p.   cm. — (Undergraduate texts in mathematics. Readings in mathematics)
    Includes bibliographical references (p.   -   ) and index.
    ISBN 978-0-387-98433-9   ISBN 978-1-4612-0523-4 (eBook)
    DOI 10.1007/978-1-4612-0523-4
    1. Mathematics—History—Sources.   I. Pengelley, David.   II. Title.
III. Series.
QA21.L34   1998
510'.9—dc21   98-3889

Printed on acid-free paper.

© 1999 Springer Science+Business Media New York
Originally published by Springer-Verlag New York in 1999
Softcover reprint of the hardcover 1st edition 1999

Production managed by Steven Pisano; manufacturing supervised by Thomas M. King.
Typeset from the authors' LaTeX files by The Bartlett Press, Inc., Marietta, GA.

9 8 7 6 5 4 3 2 (Corrected second printing, 2000)

ISBN 978-0-387-98433-9

Preface

Nothing captures the excitement of discovery as authentically as a description by the discoverers themselves. What better place to read about the search for the origin of the Nile than the report of Sir Richard Burton on his amazing journey into the depths of the African continent? The overwhelming awe and beauty of the Grand Canyon comes alive when we hear John Wesley Powell tell of his thrilling ride down its unexplored rapids. The same holds true for the explorers of the unknown territories of the mathematical world. While a second-hand account of a mathematical expedition might seem more orderly than the description of the explorer, it will likely lack the excitement, immediacy, and insights of a story told by someone who was there. It is this excitement and immediacy of mathematical discovery that we want to convey. This book contains the stories of five mathematical journeys into new realms, pieced together through the writings of the explorers themselves. Some were guided by mere curiosity and the thrill of adventure, others by more practical motives. In all cases the outcome was a vast expansion of the known mathematical world and the realization that still greater vistas remain to be explored.

In the development of calculus, one great goal was to find methods for computing areas and volumes beyond the achievements of classical Greek geometry. The potential for applications was enormous, and once success was achieved, the sciences were revolutionized. Likewise, the prospect for methods to solve general algebraic equations fostered high hopes for important applications within mathematics, as well as in the sciences and engineering. In contrast, the quest for a solution to the so-called Fermat's Last Theorem in number theory seems whimsical, apparently motivated by nothing more than the desire to meet a challenge. Nonetheless, when the goal was finally reached, mathematics had profited immeasurably from the effort. The same outcome crowned the attempt to conquer the treacherous world of infinite sets. What first seemed liked a hopeless fight against paradoxes and the tenets of theology turned into one of the greatest revolutions mathematics has ever seen. Finally, the apparently placid waters of geometry were hiding a rich world beyond the imagination of all but a handful of brave souls who dared explore the possibilities of a non-Euclidean view of geometry and the physical world. Begun out of a desire for mathematical elegance and completeness, this two-thousand-year quest led to a vast expansion of mathematics and fundamental applications to the theory of relativity. We will tell these stories as much as possible by guiding the reader through the very words of the mathematicians at the heart of these events.

This book is more about mathematics than it is about its history. Our goal is to throw light on the mathematical world we live in today, and we believe that its history is essential to understanding and appreciation. Our project began as a freshman honors course we have taught at New Mexico State University since 1989. The aim of the course is to introduce students from a wide variety of majors to the exciting world of mathematical discovery. Typically, some subsequently decide to major in mathematics. In the course we try to get across the thrill of exploring the unknown that motivates most mathematicians. Students see the mighty mountains that the community of mathematicians scales, sometimes through the joint effort of many generations. In the end a better understanding of the mathematical present is paired with the realization that mathematics is a living, breathing subject, facing new challenges every day. We hope that this book serves the same goals.

The book can be used in a variety of ways. The five chapters are completely independent of one another, as are largely the individual sections within each chapter. Necessarily, the level of difficulty within a chapter varies considerably. The chapter introduction and first sections can be appreciated and understood by someone with a good high school education in mathematics. The later sections require considerably more mathematical maturity. It is our vision that the book will be enticing both to the intellectually curious reader and to instructors and students as a course text. The introduction to each chapter summarizes the story historically and mathematically, and subsequent sections feature the original writings of major explorers in that particular story of discovery. The five introductions, together with selections from the other sections, can be used as a text for a mathematically oriented history of mathematics course. Individual chapters can be used in a serious mathematics appreciation course or as a supplement to another mathematics course. Most importantly, it is our hope that the text will encourage the creation of courses like the one from which it originated. In our one-semester course, we usually focus on just two or three chapters. There is enough material in the book for at least two semesters.

Our initial inspiration to create courses in which students learn mathematics in its historical context was William Dunham's "Great Theorems" course [44, 45]. Unlike Dunham we insist on reading primary sources. (Our one compromise is the use of English translations.) We have discussed the feasibility and the many benefits of this approach in [103, 104, 105]. The resources [24, 156, 167, 168] as well as the newsletter [90] also contain much information on using history in teaching mathematics.

Our primary sources trace five central themes in the evolution of mathematics. Our selection criteria were the importance of the source as a milestone of progress and its accessibility without extensive prior preparation. In these choices and in our own commentary we make no claim to be comprehensive in breadth, detail, or the contributions of various individuals, groups, or cultures. Ours is not a history of mathematics, but rather an exploration of some exciting mathematics through its historical artifacts.

How do we use these materials in our own teaching? Usually, we work through the introduction together with the students and jump to the later sections as the sources are mentioned. The annotation after each source is there to help with sticky points, but is used sparingly. We have included many exercises based on the original sources, and welcome more from our readers. A most useful exercise is to rewrite a source in one's own words using modern notation, filling in all the missing details.

Finally, we integrate prose readings about mathematics into the course, many from the wonderful collection [130]. We provide students with questions about these readings, and written answers then form the basis for class discussion.

We strongly encourage the reader to go beyond this book to explore the rich and rewarding world of primary sources. There are substantial collections of original sources available in English, such as [13, 14, 58, 87, 122, 160, 166]. Collected works are, of course, also a great resource [142]. We have provided many references in the text for further reading.

This book has been in the making for almost ten years. It might never have been completed without the help of many people and institutions. The directors Tom Hoeksema and Bill Eamon of the NMSU honors program provided extensive support and encouragement for the course from which this book grew. Our department heads, Carol Walker and Doug Kurtz, believed enough in our approach to teaching to help us make it into a permanent addition to our curriculum. A grant from the Division of Undergraduate Education at the National Science Foundation provided extensive resources. The NSF advisory committee, consisting of Judy Grabiner, Tom Hoeksema, and Fred Rickey, gave lots of great advice, diligent reading, and editorial suggestions on several drafts. In addition, Florence Fasanelli provided sage words of wisdom at crucial times. The grant also allowed us to involve a graduate assistant, Xenia Kramer, in the project. We owe her special thanks for her extensive contributions to research and writing, and for testing earlier drafts in the classroom as an apprentice teacher. The lion's share of the credit must go to our students, however, without whom this book would never have been written. We used early versions of the manuscript in classes at NMSU, as well as Cornell University and the two-year-long NSF-sponsored Young Scholars Mathematics Workshop in the Rockies, at Colorado College. Our students' enthusiasm convinced us that teaching with original sources can work, and their feedback greatly improved the book.

Many people have helped us in locating original sources. We could always rely on the expertise of Keith Dennis, as well as his wonderful private collection. Our research on Sophie Germain would have been impossible without the help of

Larry Bucciarelli, Catherine Goldstein, Helmut Rohlfing of the Niedersächsische Staatsbibliothek in Göttingen, and the Bibliothèque Nationale in Paris. Thanks are also due to Dave Bayer, Don Davis, and Anne-Michel Pajus. We received assistance with translations from Hélène Barcelo and Mai Gehrke. Bill Donahue, Danny Otero, Kim Plofker, and Frank Williams contributed their expertise in Latin.

A number of teachers here and elsewhere have used earlier versions of the manuscript with their students and provided many useful suggestions for improvement. We are especially grateful to Hélène Barcelo, David Arnold (and his student Charles McCoy), Danny Otero, Jamie Pommersheim, Mike Siddoway, and Irena Swanson. We thank all those who volunteered to read drafts and gave suggestions, especially Otto Bekken and Klaus Barner. We have also been most fortunate to benefit from the ideas and expertise of fellow instructors at programs where we have presented our approach and materials, including a minicourse for the Mathematical Association of America, and the NSF/MAA Institute in the History of Mathematics and its Use in Teaching, especially the detailed comments of David Dennis and Ed Sandifer.

We are very grateful to Ina Lindemann from Springer-Verlag, who showed great interest in our project and supported us with just the right mixture of patience and prodding. Her enthusiasm and great forbearance provided much encouragement. We also thank the other staff at Springer, and our copyeditor, David Kramer, for their expert assistance and excellent suggestions. Finally, we thank Rose Marquez for her expert secretarial assistance.

The second author thanks his wife, Pat Penfield, for her enduring and invaluable love, encouragement, and support for this endeavor, excellent ideas, and incisive critiques; and his parents Daphne and Ted, for their constant love, encouragement and support, inspiration, and interest in history.

October, 1998   Reinhard Laubenbacher
                                                          David Pengelley

Contents

CHAPTER 1

Geometry: The Parallel Postulate

1.1 Introduction

The first half of the nineteenth century was a time of tremendous change and upheavals all over the world. First the American and then the French revolution had eroded old power structures and political and philosophical belief systems, making way for new paradigms of social organization. The Industrial Revolution drastically changed the lives of most people in Europe and the recently formed United States of America, with the newly perfected steam locomotive as its most visible symbol of progress. The modern era began to take shape during this time (Exercise 1.1). No wonder that mathematics experienced a major revolution of its own, which also laid the foundations for the modern mathematical era. For twenty centuries one distinguished mathematician after another attempted to prove that the geometry laid out by Euclid around 300 B.C.E. in his *Elements* was the "true" and only one, and provided a description of the physical universe we live in. Not until the end of the eighteenth century did it occur to somebody that the reason for two-thousand years' worth of spectacular failure might be that it was simply *not true*. After the admission of this possibility, proof of its reality was not long in coming. However, in the end this "negative" answer left mathematics a much richer subject. Instead of one geometry, there now was a rich variety of possible geometries, which found applications in many different areas and ultimately provided the mathematical language for Einstein's relativity theory.

To understand what is meant by this statement we need to begin by taking a look at the structure and content of Euclid's *Elements*. Just like other authors before him, Euclid had produced a compendium of geometric results known at the time. What made his *Elements* different from those of his predecessors was a much higher standard of mathematical rigor, not to be surpassed until a few centuries ago, and the logical structure of the work. Beginning with a list of postulates, which we might consider as fundamental truths accepted without proof, Euclid builds up his geometrical edifice as a very beautiful and economical succession of theorems and proofs, each depending on the previous ones, with little that is superfluous. This structure was greatly influenced by the teachings of Aristotle. Naturally, much depends on one's choice of "fundamental truths" that one is willing to accept without demonstration as foundation of the whole theory.

Among the ten postulates, or axioms, as they would be called today, the five most important ones are of two types [51, vol. I, pp. 195 ff]. The first three postulates assert the possibility of certain geometric constructions.

1. To draw a straight line from any point to any point.
2. To produce a finite straight line continuously in a straight line.
3. To describe a circle with any center and radius.

The next one states that

4. All right angles are equal to one another.

Thus, a right angle is a *determinate magnitude*, by which other angles can be measured. A rather subtle consequence of this postulate is that space must be homogeneous, so that no distortion occurs as we move a right angle around to match it with other right angles. We will have more to say about this later.

Finally, the last and most important postulate concerns parallel lines. Again, faithful to Aristotelian doctrine, Euclid precedes his postulates by definitions of the concepts to be used. He defines two parallel straight lines to be "straight lines which, being in the same plane and being produced indefinitely in both directions, do not meet one another in either direction" [51, vol. I, p. 190]. The fifth, or "parallel," postulate, as it is known, states:

5. If a straight line falling on two straight lines makes the interior angles on the same side less than two right angles, the two straight lines, if produced indefinitely, will meet on that side on which are the angles less than two right angles.

It is a witness to Euclid's genius that he chose this particular statement as the basis of his geometry and viewed it as undemonstrable, as history was to show. (For more information on Euclid and his works see the chapter on number theory. A detailed description of the *Elements* can be found in [42].)

Much of what we know about the role of the *Elements* in antiquity comes from an extensive commentary by the philosopher Proclus (410–485), head of the Platonic Academy in Athens and one of the last representatives of classical Greek thought. According to him, the parallel postulate was questioned from the very beginning, and attempts were made either to prove it using the other postulates or to replace it by a more fundamental truth, possibly based on a different definition of parallelism. Proclus himself says:

This ought even to be struck out of the Postulates altogether; for it is a theorem involving many difficulties, which Ptolemy, in a certain book, set himself to solve, and it requires for the demonstration of it a number of definitions as well as theorems. And the converse of it is actually proved by Euclid himself as a theorem. It may be that some would be deceived and would think it proper to place even the assumption in question among the postulates as affording, in the lessening of the two right angles, ground for an instantaneous belief that the straight lines converge and meet. To such as these Geminus[1] correctly replied that we have learned from the very pioneers of this science not to have any regard to mere plausible imaginings when it is a question of the reasonings to be included in our geometrical doctrine. For Aristotle says that it is as justifiable to ask scientific proofs of a rhetorician as to accept mere plausibilities from a geometer; and Simmias is made by Plato to say that he recognizes as quacks those who fashion for themselves proofs from probabilities. So in this case the fact that, when the right angles are lessened, the straight lines converge is true and necessary; but the statement that, since they converge more and more as they are produced, they will sometime meet is plausible but not necessary, in the absence of some argument showing that this is true in the case of straight lines. For the fact that some lines exist which approach indefinitely, but yet remain non-secant, although it seems improbable and paradoxical, is nevertheless true and fully ascertained with regard to other species of lines. May not then the same thing be possible in the case of straight lines which happens in the case of the lines referred to? Indeed, until the statement in the Postulate is clinched by proof, the facts shown in the case of other lines may direct our imagination the opposite way. And, though the controversial arguments against the meeting of the straight lines should contain much that is surprising, is there not all the more reason why we should expel from our body of doctrine this merely plausible and unreasoned hypothesis? [51, pp. 202 f.]

This objection to the parallel postulate, so aptly described here by Proclus, was shared by mathematicians for the next two thousand years and produced a vast amount of literature filled with attempts to furnish the proof Proclus calls for. His description also gives a glimpse of the age-old debate about what constitutes mathematical rigor, which was to play an important role in the subsequent history of the problem.

Why would Euclid choose to include such an odd and unintuitive statement among his postulates? Parallels play a central role in Euclidean geometry, because they allow us to transport angles around, the central tool for proving even the most basic facts. Thus, given an angle with sides l and l', we want to draw the same angle with side l', through a point P not on l. In order to do this, we need to be able to draw a line through P that is parallel to l (Figure 1.1). This raises the question whether such parallels always exist. But to be useful for constructions there must also be a unique parallel to l through P. Euclid proves the existence of parallels in Proposition 31 of Book I, without the use of the parallel postulate. Uniqueness

---

[1] Greek mathematician, approx. 70 B.C.E.

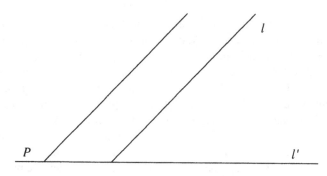

FIGURE 1.1. Transport of angles.

follows from Proposition 30 [51, Vol. I, p. 316], for which the parallel postulate is necessary (Exercise 1.2).

The first source in this chapter is Proposition 32 from Book I of the *Elements*, which asserts that the angle sum in a triangle is equal to two right angles. Book I is arranged in such a way that the first 28 propositions can all be proved without using the parallel postulate. It is used, however, in Euclid's proof of Proposition 32. Much subsequent effort was focused on understanding the precise relationship between this result and the parallel postulate, as we will see later in the chapter.

The other central consequence of the parallel postulate in Euclidean geometry is the Pythagorean Theorem,[2] perhaps the best-known mathematical result in the world. Babylonian civilizations knew and used it at least 1,000 years before Pythagoras, and from 800–600 B.C.E. the *Sulbasutras* of Indian Vedic mathematics and religion told how to use it to construct perfect religious altars [91, pp. 105, 228–229]. The earliest Chinese mathematical text in existence shows a diagrammatic proof of the theorem, based on intuitive ideas of how squares fit together (equivalent to assuming the parallel postulate). While it is difficult to date this text, it is certainly a development concurrent with, and completely separate from, classical Greek mathematics [91, pp. 132, 180][116, pp. 124, 126].

Greek mathematics continued to explore the question of the validity of the parallel postulate. Shortly after Euclid, Archimedes wrote a treatise *On Parallel Lines*, in which he replaced Euclid's definition by the property that parallel lines are those equidistant to each other everywhere [144, pp. 41 f]. The parallel postulate can then be proven, provided that one accepts as true that, for instance, a "line" equidistant to a straight line is itself a straight line. The issue was taken up again by a number of distinguished Islamic mathematicians, beginning in the ninth century. A very detailed account of all these efforts is given in [144, Ch. 2]. As knowledge of Greek and Islamic mathematics spread into Western Europe during the Renaissance, so did the desire to prove the parallel postulate. An interesting approach was proposed by the Englishman John Wallis (1616–1703). Much of the plane geometry in the *Elements* deals with similarity of triangles and other figures. Wallis gives a proof

---

[2]The sum of the squares of the legs of a right triangle equals the square of the hypotenuse.

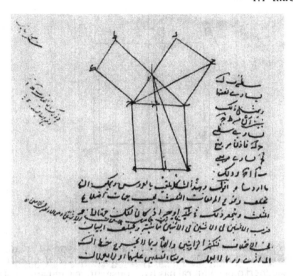

PHOTO 1.1. Pythagorean Theorem in Thabit Ibn Qurra's ninth-century translation of Euclid's *Elements*.

of the parallel postulate based on the assumption that triangles similar to a given one *exist*. Now, it is debatable whether this assumption is any more obvious than the parallel postulate itself, but Wallis's argument shows that the validity of the parallel postulate is equivalent to the possibility of shrinking or expanding figures without changing their shape [144, p. 97].

It was the Italian Jesuit Girolamo Saccheri (1667–1733) who proposed a radically new approach to the problem. It was to bring him to the brink of a revolutionary discovery, which ironically and tragically neither he nor his contemporaries realized. Rather than trying to deduce the parallel postulate from the other four, he assumed that it was false and then tried to derive a contradiction from this assumption. And so he proceeded to prove theorem after theorem of a geometry in which the parallel postulate was false, looking in vain for the hoped-for contradiction. All he used was the first 28 propositions of the *Elements*, whose proof depended only on the first four postulates. Finally, using invalid reasoning, he convinced himself that he had found the elusive contradiction, and concluded that the parallel postulate was valid after all. In 1733, he published his collection of theorems and his unfortunate conclusion in the book *Euclid Freed of All Blemish or A Geometric Endeavor in Which Are Established the Foundation Principles of Universal Geometry* [150]. After receiving quite a bit of attention upon its publication, the book was promptly forgotten for 150 years. Without realizing it, Saccheri had developed a body of theorems about a new geometry, which was free of contradictions and in which the parallel postulate was false. After two thousand years, the quest to solve the parallel problem could have come to a most surprising and wonderful resolution with the discovery that the world of geometry was much richer than humankind had realized.

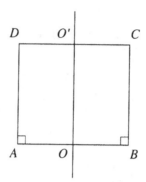

FIGURE 1.2. Saccheri's quadrilateral.

Saccheri's work centers on the nature of a special type of quadrilateral [18] (Figure 1.2). A Saccheri quadrilateral has two consecutive right angles A and B, and two equal sides AD and BC, from which one deduces (without assuming the parallel postulate) that angles C and D are equal (Exercise 1.3). Now, assuming Euclid's parallel postulate, one can prove that C and D are also right angles (Exercise 1.4). So if we assume that C and D are not right angles, we are implicitly denying the parallel postulate and replacing it by an alternative. Saccheri considered the following three possibilities:

1. The Hypothesis of the Right Angle (HRA): C and D are right angles.
2. The Hypothesis of the Obtuse Angle (HOA): C and D are obtuse angles.
3. The Hypothesis of the Acute Angle (HAA): C and D are acute angles.

Saccheri then proved several interesting and useful theorems. First, under HRA, HOA, or HAA, AB is respectively equal to, greater than, or less than CD. Second, if HRA, HOA, or HAA holds for just one Saccheri quadrilateral, then the same hypothesis holds for any Saccheri quadrilateral. Third, and most importantly for us here, under HRA, HOA, or HAA, the sum of the angles in any triangle is respectively equal to, greater than, or less than two right angles (Exercise 1.5).

Saccheri then tried to establish the parallel postulate by refuting both the HOA and HAA. (Doing so is enough to establish the parallel postulate, but we will not show the details here.) Both proofs are based on obtaining a contradiction arising from the assumed hypothesis, and while his refutation of the HOA was correct, his proof of the refutation of the HAA is based on assumptions about how lines meet at infinity and was highly questionable. While Saccheri was not the first to consider many of these connections, his work went considerably further than all before him and shows clearly that the question of the parallel postulate and its two alternatives is equivalent to the question of each of the three possibilities for the angle sum of any triangle.

Of course, Saccheri realizes that his arguments leave something to be desired.

It is well to consider here a notable difference between the foregoing refutations of the two hypotheses. For in regard to the hypothesis of obtuse angle the thing is clearer than midday light. . .

But on the contrary I do not attain to proving the falsity of the other hypothesis, that of acute angle, without previously proving that the line, all of whose points are equidistant from an assumed straight line lying in the same plane with it, is equal to [a] straight line [144, pp. 99–101].

And so he admits that to refute the HAA he had to make use of another result that he does not consider completely clear and beyond reproach. We will see that this phenomenon of recourse to other questionable and unproven assumptions is a pattern in purported proofs of the refutation of the HAA.

Not long after Saccheri's book appeared, a similar investigation was undertaken by the Swiss mathematician Johann Lambert (1728–1777), whose interest in foundational questions naturally led him to consider the parallel postulate. His treatise *Theory of Parallels* (reproduced in [50]) was not published during his lifetime, however, possibly because he felt unsatified with its inconclusiveness. His work follows that of Saccheri in its approach.

Lambert introduces his treatise with:

This work deals with the difficulty encountered in the very beginnings of geometry and which, from the time of Euclid, has been a source of discomfort for those who do not just blindly follow the teachings of others but look for a basis for their convictions and do not wish to give up the least bit of rigor found in most proofs. This difficulty immediately confronts every reader of Euclid's *Elements*, for it is concealed not in his propositions but in the axioms with which he prefaced the first book [144, pp. 99–101].

Specifically regarding the parallel postulate (which Lambert calls the "11th axiom"), he says:

Undoubtedly, this basic assertion is far less clear and obvious than the others. Not only does it naturally give the impression that it should be proved, but to some extent it makes the reader feel that he is capable of giving a proof, or that he should give it.

However, to the extent to which I understand this matter, this is just a *first* impression. He who reads Euclid further is bound to be amazed not only at the thoroughness and rigor of his proofs but also at the well-known delightful simplicity of his exposition. This being so, he will marvel all the more at the position of the 11th axiom when he finds out that Euclid proved propositions that could far more easily be left unproved.

After providing a proof that refutes the HOA, as had Saccheri, Lambert turns to the HAA. It was typical, in trying to refute the HAA, to derive consequences from it that would lead to a contradiction, thereby refuting it. In doing so, Lambert and others were in fact deriving results that would have the status of theorems in a brand new geometry in which the Hypothesis of the Acute Angle holds, replacing the parallel postulate. Lambert's point of view leans just slightly in the direction of this new geometry, rather than simply rejecting it as impossible, when he says:

> It is easy to see that under the [Hypothesis of the Acute Angle] one can go even further and that analogous, but diametrically opposite, consequences can be found under the [Hypothesis of the Obtuse Angle]. But, for the most part, I looked for such consequences under the [Hypothesis of the Acute Angle] in order to see if contradictions might not come to light. From all this it is clear that it is no easy matter to refute this hypothesis. . .

> The most striking of these consequences is that *under the [Hypothesis of the Acute Angle] we would have an absolute measure of length for every line, of area for every surface and of volume for every physical space.* This refutes an assertion that some unwisely hold to be an axiom of geometry, for until now no one has doubted that there is no absolute measure whatsoever. There is something exquisite about this consequence, something that makes one wish that the Hypothesis of the Acute Angle were true.

> In spite of this gain I would not want it to be so, for this would result in countless inconveniences. Trigonometric tables would be infinitely large, similarity and proportionality of figures would be entirely absent, no figure could be imagined in any but its absolute magnitude, astronomers would have a hard time, and so on.

Lambert had discovered that with the HAA, length was no longer relative as it is in Euclid's geometry. Specifically, one could not simply enlarge or shrink geometric figures at will, always creating similar figures with the same shape. To give a concrete example, in the new world Saccheri and Lambert are exploring while trying to refute the HAA, one finds that lengthening the sides of an equilateral triangle, say by doubling the length of each side, will reduce the size of its angles, so that the larger equilateral triangle is not similar to the smaller one. The length of its sides will determine the size of its angles, and vice versa, so that by picking a particular size (say half a right angle) for its angles, one is forcing its sides to have a certain length (which one could choose as the absolute unit of measurement in the new geometry). It is in this sense that Lambert says we would have an absolute measure of length.

How did Lambert react to this strange new world? On the one hand he found it truly enticing, on the other frightening because it seems it would be so much more complex. He recognizes, however, that his desires should not play a role:

> But all these are arguments dictated by love and hate, which must have no place either in geometry or in science as a whole.

> To come back to the [Hypothesis of the Acute Angle]. As we have just seen, under this hypothesis the sum of the three angles in every triangle is less than 180 degrees, or two right angles. But the difference up to 180 degrees increases like the area of the triangle; this can be expressed thus: if one of

two triangles has an area greater than the other then the first has an angle sum smaller than the second. . .

I will add just the following remark. Entirely analogous theorems hold under the [Hypothesis of the Obtuse Angle] except that under it the angle sum in every triangle is greater than 180 degrees. The excess is always proportional to the area of the triangle.

I think it remarkable that the [Hypothesis of the Obtuse Angle] holds if instead of a plane triangle we take a spherical one, for its angle sum is greater than 180 degrees and the excess is proportional to the area of the triangle.

What strikes me as even more remarkable is that what I have said here about spherical triangles can be proved independently of the difficulty posed by parallel lines....

Lambert is observing that although the HOA cannot hold in plane geometry, it does in fact hold in "spherical geometry," namely the geometry of the surface of a sphere, in which the "lines" are great circles on the sphere (i.e., circles whose center is the center of the sphere; these great circles provide the shortest distance between two points on the sphere and are therefore the paths preferred by airplanes when flying over the surface of the earth). Exercise 1.6 explores Lambert's claims about area and angle sum for spherical triangles. In spherical geometry some of Euclid's other postulates do not hold (for instance, there is more than one line joining pairs of diametrically opposite points, and one cannot extend a line indefinitely in length, in the sense that a great circle joins up with itself). This explains why the proofs of Saccheri and Lambert refuting the HOA do not also refute it in spherical geometry, since they rely on all of Euclid's other postulates, some of which are missing in spherical geometry.

Lambert speculates that:

From this I should almost conclude that the [Hypothesis of the Acute Angle] holds on some imaginary sphere. At least there must be something that accounts for the fact that, unlike the [Hypothesis of the Obtuse Angle], it has for so long resisted refutation on planes.

Lambert had reason to think that if the HOA holds on an ordinary sphere, then the HAA, which is its opposite, might hold on a sphere of imaginary radius. This idea was actually not so far-fetched, and would be made more precise later by others. Despite his entrancement with the possibilities of this new geometry on an imaginary sphere, based on the HAA, Lambert still felt he should, and could, disprove the HAA for plane geometry, and he provided his own proof, which, like all those before him, was spurious in its own way.

The last serious attempt to prove the parallel postulate was made by the French school, which dominated mathematics at the end of the eighteenth and beginning of the nineteenth century. Here, too, no thought was given to the possibility that the parallel postulate may be an assumption independent of the rest of geometry. An important argument that would have stifled any such doubts came from physics, put forward by Laplace (1749–1827), and described in [18, pp. 53 f.]:

> Laplace observes that Newton's Law of Gravitation, by its simplicity, by its generality and by the confirmation which it finds in the phenomena of nature, must be regarded as rigorous. He then points out that one of its most remarkable properties is that, if the dimensions of all the bodies of the universe, their distances from each other, and their velocities, were to decrease proportionally, the heavenly bodies would describe curves exactly similar to those which they now describe, so that the universe, reduced step by step to the smallest imaginable space, would always present the same phenomena to its observers. These phenomena, he continues, are independent of the dimensions of the universe, so that the simplicity of the laws of nature only allows the observer to recognize their ratios. Referring again to this astronomical conception of space, he adds in a Note: "The attempts of geometers to prove Euclid's Postulate on Parallels have been up till now futile. However, no one can doubt this postulate and the theorems which Euclid deduced from it. Thus the notion of space includes a special property, self-evident, without which the properties of parallels cannot be rigorously established. The idea of a bounded region, e.g., a circle, contains nothing which depends on its absolute magnitude. But if we imagine its radius to diminish, we are brought without fail to the diminution in the same ratio of its circumference and the sides of all the inscribed figures. This proportionality appears to me a more natural postulate than that of Euclid, and it is worthy of note that it is discovered afresh in the results of the theory of universal gravitation."

Laplace, like Wallis, is observing that if we allow for the existence of similarity, then Euclid's theory of parallels follows. But Laplace believed that this similarity is inherent in the physical laws of space.

A particularly clear and very instructive proof of the parallel postulate was given by Adrien-Marie Legendre (1752–1833), an important and influential member of the French Academy of Sciences, in Paris. We include it as the second source of this chapter, representing the very end of the long string of such proofs. It is taken from Legendre's textbook *Eléments de Géométrie* (Elements of Geometry) [108], first published in 1794. The book went through many editions and was an influential geometry text all through the nineteenth century. Just like Saccheri and Lambert before him, he first refutes HOA, and then proposes a proof to refute HAA. In it he uses an interesting, apparently completely obvious assumption, which, however, turns out also to be equivalent to assuming the parallel postulate.

In Laplace's argument above we see one of the most important reasons why so many brilliant mathematicians over so many centuries stubbornly clung to the belief that Euclid's parallel postulate should follow from the others. Geometry was inextricably tied to space, our physical universe. And space was considered infinite, homogeneous, and the basis for all our experience. Nothing other than

Euclidean geometry was *thinkable*. Another, more subtle, reason is suggested in the preface to a modern reprint of Saccheri's book.

> At the present day, we have an abundance of organized knowledge, which offers explanations—in which we have the fullest confidence—of many aspects of our physical universe. We need only refer to thermodynamics, geophysics, fluid dynamics, paleontology—the list is endless, and no one can possibly master all the knowledge that is available. But none of this knowledge extends back more than two hundred years. A group of educated eighteenth-century men, for example, sitting before an open fire, could no more understand or explain the nature of that fire than could their Neanderthal ancestors. For, the nature of light, the nature of heat, the nature of chemical combination and, in particular, of combustion, even the existence of oxygen, were yet to be discovered. Nor did Science, in the eighteenth century, at all inspire confidence, as it does today...
>
> Let us therefore take a brief inventory of what existed in the eighteenth century to satisfy man's craving for certainty, for organized knowledge.
>
> Physical science, as we have just mentioned, was not yet ready to satisfy this need for certainty. The teachings of the Church were indeed unquestioned Truths, for the Faithful. But the Faithful had to be aware that these Truths were ignored by much of mankind and were under constant attack by heretics. Philosophy seemed to offer certainty, but the existence of competing, and contradictory, schools of philosophy betrayed an underlying uncertainty.
>
> Contrast all of this with Geometry, which for two thousand years had been accepted as being the Science of the space in which we live...
>
> If a valid geometry, alternative to Euclid's, were to exist, then Euclidean geometry would not necessarily be the science of space, and in fact there would no longer *be* a science of space. And with that science gone, there would be nothing—no science at all. Thus, in addition to the many reasons for not doubting Euclidean geometry to be the one and only geometry (after all, in our day, it is still the geometry of architecture, engineering, and most branches of the sciences), we have another and powerful *silent* motive—a motive which does not reach consciousness and which for that reason is all the more powerful, the sort of motive which, under the right circumstances, makes an idea *unthinkable*. [150, pp. ix–x]

What makes this seemingly stubborn pursuit of the parallel postulate all the more puzzling is that during this entire time a perfectly good non-Euclidean geometry was sitting right under people's noses: the geometry of the sphere. But since Euclidean geometry was tied so strongly to the nature of space itself, the step of viewing plane and spherical geometry as just two examples of geometrical systems of equal status never suggested itself.

PHOTO 1.2. Gauss.

But the time was finally ripe for a breakthrough. As in so many other branches of mathematics it was left to Carl Friedrich Gauss (1777–1855), the mathematical titan of the nineteenth century, to make the first step. Ironically, for fear of getting embroiled in controversy, he kept his insights secret for almost fifty years, until others had taken the courageous step to proclaim the existence of a geometry independent of the parallel postulate.

Beginning in 1792, Gauss at first also tried to prove the postulate, like his predecessors, proceeding by assuming it to be false, hoping to reach a contradiction. In a letter to his fellow student Wolfgang Bolyai (1775–1856), who was also working on this problem and had convinced himself of success, he expresses his frustration:

> As for me, I have already made some progress in my work. However, the path I have chosen does not lead at all to the goal which we seek, and which you assure me you have reached. It seems rather to compel me to doubt the truth of geometry itself.

> It is true that I have come upon much which by most people would be held to constitute a proof: but in my eyes it proves as good as *nothing*. For example, if one could show that a rectilinear triangle is possible, whose area would be greater than any given area, then I would be ready to prove the whole of geometry absolutely rigorously.

Most people would certainly let this stand as an Axiom; but I, no! It would, indeed, be possible that the area might always remain below a certain limit, however far apart the three angular points of the triangle were taken [18, pp. 65 f.].

Some time later Gauss finally convinced himself that the right path was in fact to give up this age-old attempt and instead develop a new geometry, which he called *Anti-Euclidean* and later *Non-Euclidean*. In a letter to his friend and colleague Heinrich Olbers (1758–1840), Gauss writes:

I am ever more convinced that the necessity of our geometry cannot be proved—at least not by *human* reason for human reason. It is possible that in another lifetime we will arrive at other conclusions on the nature of space that we now have no access to. In the meantime we must not put geometry on a par with arithmetic that exists purely a priori but rather with mechanics [144, p. 215].

Thus, Gauss too had been hampered by the dominant philosophy of space and its geometry, expressed earlier by Laplace, and championed by Immanuel Kant (1724–1804), the most influential philosopher of the eighteenth century. (See [144] for a detailed discussion of the influence of philosophies of space on the parallel problem.)

Gauss bases everything on the following definition of parallel lines (Figure 1.3). *If the coplanar straight lines AM, BN do not intersect each other, while on the other hand, every straight line through A between AM and AB cuts BN, then AM is said to be parallel to BN* [18, pp. 67 f.].

The lines beginning at A and extending to the right are divided into two classes, those that intersect BN and those that do not. If we think of these lines as generated by taking the line AB extended upwards and rotating it around A clockwise, then the first line that does not intersect BN is called *parallel* to BN (Exercise 1.7).

Without the assumption of the parallel postulate, there could be more than one line that does not intersect BN. From this definition Gauss proceeds to prove the fundamental theorems of a new geometry. But he chose to keep his discoveries to himself, except for some close friends. When Wolfgang Bolyai sent him a paper that Bolyai's son János had written, in which he proposes just such a new geometry, Gauss replies:

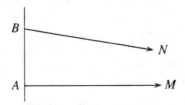

FIGURE 1.3. Gauss's definition.

If I commenced by saying that I must not praise this work you would certainly be surprised for a moment. But I cannot say otherwise. To praise it, would be to praise myself. Indeed the whole contents of the work, the path taken by your son, the results to which he is led, coincide almost entirely with my meditations, which have occupied my mind partly for the last thirty or thirty-five years. So I remained quite stupefied. So far as my own work is concerned, of which up till now I have put little on paper, my intention was not to let it be published during my lifetime. Indeed the majority of people have no clear ideas upon the questions of which we are speaking, and I have found very few people who could regard with any special interest what I communicated to them on this subject. To be able to take such an interest it is first of all necessary to have developed careful thought to the real nature of what is wanted and upon this matter almost all are most uncertain. On the other hand, it was my idea to write down all this later so that at least it should not perish with me. It is therefore a pleasant surprise for me that I am spared this trouble, and I am very glad that it is just the son of my old friend who takes the precedence of me in such a remarkable manner [144, pp. 215–217].

János Bolyai (1802–1860) was a Hungarian officer in the Austrian army and inherited his interest in mathematics in general and in the parallel postulate in particular from his father. In 1823, he wrote to his father:

I have now resolved to publish a work on parallels. . . I have not yet completed the work, but the road that I have followed has made it almost certain that the goal will be attained, if that is at all possible: the goal is not yet reached, but I have made such wonderful discoveries that I have been almost overwhelmed by them, and it would be the cause of continual regret if they were lost. When you see them, you too will recognize them. In the meantime I can say only this: *I have created a new universe from nothing.* All that I have sent you till now is but a house of cards compared to a tower. I am as fully persuaded that it will bring me honour, as if I had already completed the discovery [78, p. 107].

His father replied with excitement:

[I]f you have really succeeded in the question, it is right that no time be lost in making it public, for two reasons: first, because ideas pass easily from one to another, who can anticipate its publication; and secondly, there is some truth in this, that many things have an epoch, in which they are found in several places, just as violets appear on every side in the Spring. Also every scientific struggle is just a serious war, in which I cannot say when peace will arrive. Thus we ought to conquer when we are able, since the advantage is always to the first comer [78, p. 107].

Wolfgang Bolyai agreed to publish his son's manuscript as an appendix to his own book *Tentamen*, on the foundations of several mathematical subjects including geometry, which appeared in 1831. After sending a copy to Gauss, he received the above-mentioned reply. Gauss's letter dealt a devastating blow to János, crippling his whole subsequent career. While he continued to work on mathematics, he never again published anything. The lack of attention that his manuscript received in subsequent years caused only further discouragement.

There were indeed violets appearing in other places. At Kazan University, in Russia, the mathematics professor Nikolai Lobachevsky (1792–1856) wrote his first major work on geometry in 1823. Subsequent research led him down the exact same path as Gauss and Bolyai had followed. His new geometry, which was very similar to theirs, appeared first in his article *On the Principles of Geometry*, published in 1829–30 in the journal *Kazan Messenger*, produced by Kazan University. In 1835 he published a longer article on his new geometry, which also appeared in French translation in *Journal für die reine und angewandte Mathematik*, one of the foremost European mathematical journals. Then, in 1842 he published the book *Geometrische Untersuchungen zur Theorie der Parallellinien* (Geometrical Researches on the Theory of Parallels), in German, which finally brought recognition to his accomplishment. On the recommendation of Gauss, he was elected to the Göttingen Science Society, a considerable honor. The next and major source in this chapter is a part of Lobachevsky's book. He begins with a definition of parallel lines quite similar to that of Gauss, described earlier, and proceeds to derive all the fundamental theorems of his geometry without the assumption of the parallel postulate. Most importantly, he derives all the basic trigonometric formulas valid in Lobachevskian geometry, including those for triangles.

Thus, the credit for the discovery of hyperbolic geometry, as it is now known, fell in large part to Lobachevsky, who was the first to publish his account. But ultimately he did not fare much better than Bolyai. Lobachevsky being relatively unknown, his work did not receive the instant attention it deserved, and acceptance of Lobachevskian geometry was rather slow in coming. Eventually, three factors led to widespread acceptance of the brave new world of non-Euclidean geometry that was opened up by the pioneers Gauss, Bolyai, and Lobachevsky.

First of all, after Gauss's death, his correspondence with his colleague H.K. Schumacher was published between 1860 and 1863. It makes abundantly clear that Gauss thought very highly of the work of both Bolyai and Lobachevsky. His approval of these two until then unknown mathematicians did much to lend credibility to their work [18, pp. 122 ff.].

Secondly, the visionary work of Bernhard Riemann (1826–1866), one of the most imaginative mathematicians ever, proposed an entirely new paradigm for the concept of space and geometry, with a natural place for hyperbolic geometry. In his legendary paper *On the Hypotheses Which Lie at the Foundations of Geometry* [139] Riemann proposed the study of curved surfaces and higher-dimensional spaces, such as the plane or the sphere. His profound insight was that the geometry that is valid on the surface depends on its "curvature," measured by a certain constant K that is intrinsic to the surface. This idea was initially proposed by Gauss and represented a radical departure from the conventional idea that the curvature of a surface made sense only when viewed within an ambient space.

PHOTO 1.3. Riemann.

Given two points on the surface, the shortest curve that connects the two points will then play the role of a straight line between two points in the plane. Such curves are now called geodesics. For instance, for the plane, the curvature constant K is zero, and the resulting geometry is just the Euclidean one. The same is true for a surface that can be deformed, without stretching, into the plane, such as a part of a cylinder. For a sphere, on the other hand, we obtain that K is positive; that is, the sphere has positive curvature. The resulting geometry is, of course, spherical geometry, and the HOA is valid. Likewise, a surface for which any part can be deformed into part of a sphere supports the same type of geometry. This leaves surfaces with negative curvature. For those that have so-called constant negative curvature, the geometry that is valid for them is precisely hyperbolic geometry and the HAA [18, pp. 130 ff.][78, Ch. 12].

As Riemann's ideas became accepted, so did hyperbolic geometry. The result was a whole new mathematical theory: differential geometry and topology. When Albert Einstein was struggling with the theory of special and general relativity, it was this theory that provided the natural language for it. Ultimately, it turned out that the physical universe we live in looks a lot more like Lobachevsky's world than Euclid's.

Thirdly, and most importantly, the issue that needed to be settled before hyperbolic geometry was put on a firm foundation was the question of true independence of the parallel postulate from the other four. Lobachevsky had proven many theorems based on the assumption that the parallel postulate was false, without encountering a contradiction. But that did not necessarily imply that there wasn't one to be found someplace else. What was needed was a rigorous proof that the existence of more than one line through a given point parallel to a given line led to a consistent geometric theory. The ingenious solution to this problem was to produce a Euclidean model of hyperbolic geometry, a sort of faithful projection of the hyperbolic plane onto part of a Euclidean plane, in such a way that parallels, triangles, etc. in hyperbolic geometry corresponded to some type of figure in Euclidean geometry. Then, if a contradiction existed in hyperbolic geometry, it would also have to exist in Euclidean geometry. Several such models were proposed, beginning with those of the Italian mathematician Eugenio Beltrami (1835–1900) [164, p. 35][78, Ch. 13]. As the last source in this chapter we study a model given by the great French mathematician Henri Poincaré (1854–1912). He was led to this model almost incidentally through his work in the analysis of functions. In his model, the hyperbolic plane is represented as a Euclidean disk. That is, points in the hyperbolic plane correspond to Euclidean points inside the disk, and lines in the hyperbolic plane correspond to arcs of Euclidean circles meeting the boundary of the disk perpendicularly, or to diameters of the disk.

Thus, the issue of the consistency of hyperbolic geometry was now squarely in the court of Euclidean geometry. Two thousand years of scrutiny had revealed many cracks in Euclid's foundations and many weaknesses of proofs in the *Elements*. For one, Euclid's definitions were for the most part wholly inadequate. For instance, to define a point as "that which has no part" [51, p. 153] immediately calls out for a definition of "part" and so on, leading to an infinite chain of definitions. In 1899, the German mathematician David Hilbert (1862–1943), on his way to becoming the dominating figure in mathematics during the first quarter of the twentieth century, presented a new system of axioms and definitions for Euclidean geometry [88]. Euclid's five axioms are replaced by a much longer list, and notions like "point" and "line" remained undefined. Only their mutual relationships were specified.

The transition had been made from viewing geometry as the science of the space we live in to geometry as a system of axioms, which is valid as long as the axioms do not lead to contradictory results. All over mathematics the so-called axiomatic method took hold, as evidenced, for instance, in the set theory chapter of this book, as the distinctive mark of twentieth-century mathematics. The failed attempt to prove that Euclid's was the one and only geometry led to a vast new mathematical universe, which is today continuing to enrich mathematics as well as its connections to the other sciences.

Exercise 1.1: Read about world history from 1750 until 1850.

Exercise 1.2: Look up the proof of Proposition 30 in Book I of the *Elements* [51, vol. I, pp. 316–317] and identify the steps that require the parallel postulate.

Exercise 1.3: Prove that angles C and D are equal in Saccheri's quadrilateral. (Hint: Look at Book I of the *Elements*.)

Exercise 1.4: Prove that with the parallel postulate, C and D are right angles in Saccheri's quadrilateral. (Hint: Use Euclid's Proposition 27.)

Exercise 1.5: Prove Saccheri's theorems on his quadrilaterals.

Exercise 1.6: Explore what triangles look like in spherical geometry (draw pictures). How long can lines be? How small or large can the angle sum of a triangle be? Give examples. Explore how the area of a spherical triangle appears to be related to its angle sum.

Exercise 1.7: Convince yourself that Gauss's definition of parallel lines does not depend on the choice of the points A and B on the given lines.

1.2 Euclid's Parallel Postulate

Euclid's *Elements* [51], written in thirteen "books" around 300 B.C.E., compiled much of the mathematics known to the classical Greek world at that time. It displayed new standards of rigor in mathematics, proving everything by proceeding from the known to the unknown, a method called synthesis. And it ensured that a geometrical way of viewing and proving things would dominate mathematics for two thousand years.

PHOTO 1.4. Euclid.

Book I contains familiar plane geometry, Book II some basic algebra viewed geometrically, and Books III and IV are about circles. Book V, on proportions, enables Euclid to work with magnitudes of arbitrary length, not just whole number ratios based on a fixed unit length. Book VI uses proportions to study areas of basic plane figures. Books VII, VIII, and IX are arithmetical, dealing with many aspects of whole numbers, such as prime numbers, factorization, and geometric progressions. Book X deals with irrational magnitudes. The final three books of the *Elements* study solid geometry. Book XI is about parallelepipeds, Book XII uses the method of exhaustion to study areas and volumes for circles, cones, and spheres, and Book XIII constructs the five Platonic solids inside a sphere: the pyramid, octahedron, cube, icosahedron, and dodecahedron [42]. For more information about Euclid and his environment, see the number theory chapter.

Our focus will be specifically on the controversial parallel postulate in Book I and its ramifications, in particular the angle sum of triangles. Book I begins with twenty-three "definitions," five "postulates," and five "common notions." The first few definitions are fundamental ones like "A *point* is that which has no part," "A *line* is breadthless length," and "A *straight line* is a line that lies evenly with the points on itself." (Notice that a line is thus what we would call a curve, and that certain lines are then called straight.) While at first sight these definitions of the most basic objects in geometry appear to provide a good foundation, in fact they ultimately raise more questions than they answer. To use them we must first be able to work with terms like "part," "breadthless," "length," "evenly with," and we find that attempting to define these terms leads us even further backwards to yet more undefined notions. This process of defining the most basic terms leads to an infinite regression, and even begins to lead us in circles. So a twentieth-century underpinning for geometry (and other fields of mathematics as well; see the set theory chapter) follows a different approach [88], instead taking words like *point* and *line* as undefined terms, about which one simply supposes basic properties that dictate how they interact with each other. For Euclid, these assumed interactions are provided by his five postulates, already given in the Introduction. For instance, the first postulate, "To draw a straight line from any point to any point," dictates that two points always determine a line. Euclid's five common notions are axioms of a more general nature, not confined to geometry, but rather assumptions common to all the sciences as a method of arguing. For example, the first states "Things that are equal to the same thing are also equal to one another."

Euclid first develops as much as he can of plane geometry without using the parallel postulate (Postulate 5). The first 26 propositions of Book I deal primarily with familiar properties of lines, angles, and triangles. Even when he begins the theory of parallels, Propositions 27 and 28 first prove what results he could deduce about parallels without appealing to Postulate 5. While much of Euclid's *Elements* is a feat of organizing already existing knowledge into a systematic format of presentation and proof, the approach to the theory of parallels is apparently due to Euclid himself, particularly the genius of recognizing and adopting the parallel postulate as an unprovable assumption on which to base the theory [42]. To consider his theory carefully, and obtain his result on angle sums in triangles, we begin with his definitions of right angles and parallels, and restate the parallel postulate itself.

Euclid, from
Elements

BOOK I.
DEFINITION 10.

When a straight line set up on a straight line makes the adjacent angles equal to one another, each of the equal angles is right, and the straight line standing on the other is called a perpendicular to that on which it stands.

Note that this definition makes no reference to any units of angle measurement, as opposed to a definition such as "A right angle has measure 90 degrees." Of course, once the notion of degree measure of angles is introduced, Euclid's definition is equivalent to the "measure is 90 degrees" definition. (Why?) One might think of Euclid's use of "equal" above as geometric congruence. Drawing the figure described in the definition (Figure 1.4), we see that angles *ACD* and *DCB* are equal if they lie on top of each other when we fold along *DC*.

DEFINITION 23.

Parallel straight lines are straight lines which, being in the same plane and being produced indefinitely in both directions, do not meet one another in either direction.

POSTULATE 5.

That, if a straight line falling on two straight lines make the interior angles on the same side less than two right angles, the two straight lines, if produced indefinitely, meet on that side on which are the angles less than the two right angles.

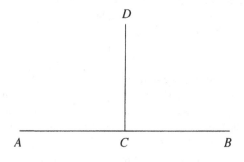

FIGURE 1.4. Definition 10.

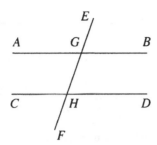

FIGURE 1.5. Proposition 29.

PROPOSITION 29.

A straight line falling on parallel straight lines makes the alternate angles equal to one another, the exterior angle equal to the interior and opposite angle, and the interior angles on the same side equal to two right angles.

This is the first result in *Elements* whose proof requires Postulate 5. We leave the proof to the reader (Exercise 1.9). In Figure 1.5, alternate angles are illustrated by the pair *AGH, GHD*, exterior and interior opposite angles are the pair *EGB, GHD*, and interior angles on the same side are *BGH, GHD*. The converse of the first claim (about alternate angles) in Proposition 29 is also true, and Euclid already proved it as Proposition 27.[3] It is interesting that this converse does not depend on Postulate 5 (Exercise 1.10).

PROPOSITION 30.

Straight lines parallel to the same straight line are also parallel to one another.

PROPOSITION 31.

Through a given point to draw a straight line parallel to a given straight line.

Let *A* be the given point, and *BC* the given straight line; thus it is required to draw through the point *A* a straight line parallel to the straight line *BC*. [See Figure 1.6.]

Let a point *D* be taken at random on *BC*, and let *AD* be joined; on the straight line *DA*, and at the point *A* on it, let the angle *DAE* be constructed equal to the angle *ADC* [I. 23]; and let the straight line *AF* be produced in a straight line with *EA*.

Then, since the straight line *AD* falling on the two straight lines *BC*, *EF* has made the alternate angles *EAD*, *ADC* equal to one another, therefore *EAF* is parallel to *BC* [I. 27].

---

[3]The full statement of Proposition 27 is "If a straight line falling on two straight lines make the alternate angles equal to one another, the straight lines will be parallel to one another."

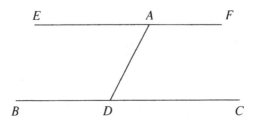

FIGURE 1.6. Proposition 31.

Therefore, through the given point A the straight line EAF has been drawn parallel to the given straight line BC.

In the proof, when Euclid makes use of a previously proven proposition, he refers to it in brackets.

Notice that Proposition 31 does not depend on the parallel postulate. In wondering why Euclid did not therefore state and prove it earlier, when he was proving everything he possibly could without the postulate, Proclus suggests the following [51, Vol. I, p. 316]. Proposition 31 implies that there is only one straight line through A parallel to BC, but does not actually prove this. However, this fact will follow from Proposition 30 (whose proof, which we have omitted, does require the parallel postulate). See Exercises 1.11 and 1.12.

Now we are ready for Euclid's theorem on the angle sum of triangles.

PROPOSITION 32.

In any triangle, if one of the sides be produced, the exterior angle is equal to the two interior and opposite angles, and the three interior angles of the triangle are equal to two right angles.

Let ABC be a triangle, and let one side of it BC be produced to D [see Figure 1.7]
I say that the exterior angle ACD is equal to the two interior and opposite angles CAB, ABC, and the three interior angles of the triangle ABC, BCA, CAB are equal to two right angles.

For let CE be drawn through the point C parallel to the straight line AB [I. 31].

Then, since AB is parallel to CE, and AC has fallen upon them, the alternate angles BAC, ACE are equal to one another [I. 29].

Again, since AB is parallel to CE, and the straight line BD has fallen upon them, the exterior angle ECD is equal to the interior and opposite angle ABC [I. 29].

But the angle ACE was also proved equal to the angle BAC;
therefore, the whole angle ACD is equal to the two interior and opposite angles BAC, ABC.

Let the angle ACB be added to each;
therefore, the angles ACD, ACB are equal to the three angles ABC, BCA, CAB.

But the angles ACD, ACB, are equal to two right angles [I.13];
therefore, the angles ABC, BCA, CAB are also equal to two right angles.

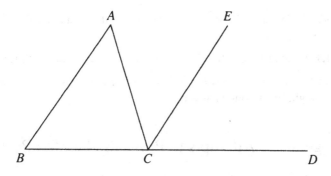

FIGURE 1.7. Proposition 32.

. . .

Euclid has proved, using the parallel postulate, that the angle sum in a triangle is always two right angles. In fact, as we discussed in the Introduction, this property of triangles is equivalent to the parallel postulate; i.e., one can also prove the converse implication, that if the angle sum is assumed to be two right angles, then the parallel postulate follows (without assuming it in advance). Thus proving the parallel postulate is equivalent to proving the angle sum theorem. Many later efforts at proving the parallel postulate homed in on precisely this approach, as we will see in our next source.

Exercise 1.8: Prove Proposition 15: "If two straight lines cut one another, they make the vertical angles equal to one another."

In Figure 1.8, the angles *AED* and *CEB* form one pair of vertical angles, and the angles *AEC* and *BED* form another. (The term "vertical" refers to the shared vertex *E*; as Heath notes in his commentary in [51, Vol. I, p. 278], "vertically opposite" is clearer.) Then compare your proof with Euclid's.

Exercise 1.9: Prove Proposition 29. Along the way you may find that you prove some other propositions of Euclid's (see, for example, Exercise 1.8). Then compare your proof with Euclid's.

Exercise 1.10: Prove Proposition 27: "If a straight line falling on two straight lines make the alternate angles equal to one another, the straight lines will be parallel

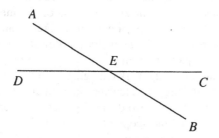

FIGURE 1.8. Proposition 15.

to one another." You should not need to use the parallel postulate. Then compare your proof with Euclid's.

Exercise 1.11: Prove Proposition 30. Then compare your proof with Euclid's.

Exercise 1.12: Use Propositions 30 and 31 to show that there is exactly one parallel to a given straight line through a given point.

1.3   Legendre's Attempts to Prove the Parallel Postulate

Adrien-Marie Legendre was born to a well-to-do family, and received an excellent education at schools in Paris. His family's modest fortune allowed him to devote himself entirely to research, although he did teach mathematics for a time at the Ecole Militaire in Paris. In 1782 Legendre won a prize from the Berlin Academy for his essay on the subject "Determine the curve described by cannonballs and bombs, taking into consideration the resistance of the air; give rules for obtaining the ranges corresponding to different initial velocities and to different angles of projection." Then, as today, political and military considerations had great influence on the directions of mathematical research. This prize, along with work on celestial mechanics, gained Legendre election to the French Academy of Sciences in 1783, and his scientific output continued to grow. Legendre's favorite areas of research were celestial mechanics, number theory (see the number theory chapter), and the theory of elliptic functions, of which he should be considered the founder.

From 1787 Legendre was heavily involved in the Academy's work with the geodetic research of the Paris and Greenwich observatories, such as the linking of their meridians, and this stimulated him to work on spherical geometry, where he is known for his theorem on spherical triangles. Thus he already had a very practical knowledge and interest in geometry, and an understanding of the differences between plane and spherical geometry, before he worked on the parallel postulate.

During the ravages caused by the French Revolution, Legendre lost his small fortune. However, in 1791 he became one of the commissioners for the astronomical operations and triangulations necessary for determining the meter and establishing our worldwide metric system of measurements. Soon thereafter he became head of part of the National Executive Commission of Public Instruction, and the Committee on Public Instruction commissioned him to write what became his big textbook *Eléments de Géométrie* (Elements of Geometry) [108]. It first appeared in 1794, and many subsequent editions in French and other languages dominated elementary instruction in geometry for almost a century. At this time Legendre also supervised part of an enormous project to prepare new tables of functions needed for the surveying work emerging from the adoption of the metric system. For instance, the project calculated sines of angles in ten-thousandths of a right angle, correct to twenty-two decimal places. It is hard for us today, in the computer age, to imagine the scope of such an enterprise. Legendre wrote "These... tables, constructed by means of new techniques based principally on

PHOTO 1.5. Legendre.

the calculus of differences, are one of the most beautiful monuments ever erected to science." Legendre arranged to prepare the tables by two completely different methods, one based on new formulas of his for successive differences of sines. The other method involved only additions, and could be done by people with few mathematical skills. Then the two independently obtained tables were compared to verify their correctness. In 1813 Legendre succeeded Lagrange (see the analysis and number theory chapters) at the Bureau des Longitudes (Office of Longitudes), a post he retained for the rest of his life [42].

As we have seen in the Introduction, Legendre's work on the parallel postulate followed a long line of geometers over many centuries who tried to resolve its status, often attempting to prove it or claiming to have done so [18, 144]. By the early eighteenth century, endeavors to resolve the parallel postulate had focused in particular directions, exemplified in the work of Saccheri and Lambert, and particularly its relationship to the angle sum in triangles [18].

Reading Legendre will be illuminating in several ways. Legendre was the last important mathematician to firmly believe that Euclid's geometry, based on the parallel postulate, was the only possible one. He was convinced that the parallel postulate held true, and that he had a variety of valid proofs for it, often based on proving the equivalent result that the angle sum of any triangle must be two right angles. However, he struggled unsuccessfully for almost forty years to convince others of this, and to find a proof that he considered simple and straightforward enough to present to students in his textbook.

In almost every edition of his *Eléments*, Legendre proved the parallel postulate. However, each proof was attacked as insufficient by other mathematicians. With astonishing and stubborn persistence, refusing to consider the possibility that the parallel postulate might be false, he would provide a new proof in a later edition, hoping to satisfy his critics. In the third edition (1800) he replaced his original proof with the one we will examine, oriented towards the angle sum in a triangle. In the ninth edition (1812) he abandoned this proof, returning, as he later explained, "to Euclid's simple way of proceeding, referring to notes [appendices] for the rigorous demonstration" [108, 12th ed. (1823), Note II, p. 279]. He was particularly concerned with finding a proof suitable to insert in his *Eléments* for students to learn, and in the twelfth edition (1823) he believed he had discovered the right one, which he kept in all the remaining editions published during his lifetime. (It was flawed, too, of course.)

Despite the fact that others pointed out the flaw in each of his proofs, any doubts Legendre had about his arguments, and about the veracity of the parallel postulate, were only about pedagogical suitability for students, not about the truth of his claims. In a large memoir he finished in 1832, less than a year before he died, entitled *Reflections on the Different Ways of Proving the Theory of Parallels or the Theorem on the Sum of the Three Angles of the Triangle*, he wrote:

> These considerations. . . leave little hope of obtaining a proof of the theory of parallels or the theorem on the sum of the three angles of the triangle, by means as simple as those which one uses for proving the other propositions of the *Eléments*.

> It is no less certain that the theorem on the sum of the three angles of the triangle must be regarded as one of those fundamental truths which is impossible to dispute, and which are an enduring example of mathematical certitude, which one continually pursues and which one obtains only with great difficulty in the other branches of human knowledge. Without doubt one must attribute to the imperfection of common language and to the difficulty of giving a good definition of the straight line the little success which geometers have had until now when they have wanted to deduce this theorem only from the notions of equality of triangles contained in the first book of the *Eléments*[4] [109, pp. 371–372].

Stubbornly refusing to doubt the truth of the postulate, he even explained that an important reason for writing this final big memoir was that the proof he had given in the twelfth and all subsequent editions "had not gained the agreement of certain professors who, without disputing its accuracy, had found it too difficult for their students to understand. . . " [109, p. 407].

---

[4] of Euclid.

Let us now examine the proof Legendre offers in the third (1800) through eighth editions [108, 5th ed. (1804), Book I, pp. 19–22, plate 2]. First he proves the rejection of the Hypothesis of the Obtuse Angle (as had Saccheri) by proving that triangles cannot have angle sum exceeding two right angles. Then he finishes the other half, "proving" the rejection of the Hypothesis of the Acute Angle (just as Saccheri had) by showing that triangles cannot have angle sum less than two right angles. Thus the only possibility left is that all triangles have angle sum equal to two right angles, which is equivalent to the parallel postulate. See if you can find the hidden unsupported assumption which irremediably mars his proof! (See Exercise 1.13.)

We have bracketed in the text Legendre's marginal references to previous results used, showing how he followed the model of Euclid's *Elements*.

Legendre, from
Elements of Geometry

PROPOSITION XIX.
LEMMA.

The sum of the three angles of a triangle cannot be larger than two right angles.

Suppose, if it is possible, that ABC is a triangle in which the sum of the three angles is larger than two right angles [Fig. 1.9].

On the extension of AC take $CE = AC$; make the angle $ECD = CAB$, the side $CD = AB$, join DE and BD. The triangle CDE will be equal to triangle BAC, because they have an angle contained between corresponding equal sides [pr. 6]; thus one will have angle $CED = ACB$, angle $CDE = ABC$, and the third side ED equal to the third BC.

Because the line ACE is straight, the sum of the angles ACB, BCD, DCE is equal to two right angles [pr. 2, cor. 3]; now one supposes the sum of the angles of triangle ABC greater than two right angles; one thus has

$$CAB + ABC + BCA > ACB + BCD + ECD;$$

subtracting from each side the common ACB and $CAB = ECD$, there remains $ABC > BCD$; and because the sides AB, BC of triangle ABC, are equal to the sides CD, CB of triangle BCD, it follows that the third side AC is larger than the third BD [pr. 10].

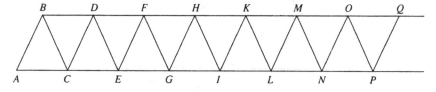

FIGURE 1.9. Legendre's refutation of HOA.

Let us imagine now that one extends the line AC indefinitely, likewise the sequence of equal and similarly placed triangles ABC, CDE, EFG, GHI, etc.; if one joins neighboring vertices with the lines BD, DF, FH, HK, etc., one will form at the same time a sequence of intermediate triangles BCD, DEF, FGH, etc., which will all be equal to each other, because they will have an equal angle contained between corresponding equal sides; thus one will have $BD = DF = FH = HK$, etc.

This stated, because one has $AC > BD$, letting the difference $AC - BD = D$, it is clear that $2D$ will be the difference between the straight line ACE equal to $2AC$ and the straight or broken line BDF equal to $2BD$; so that one has $AE - BF = 2D$. One has similarly $AG - BH = 3D$, $AI - BK = 4D$, and so forth. But, no matter how small the difference D may be, it is clear that this difference, repeated sufficiently many times, will have to be greater than a given length. One can thus suppose that the sequence of triangles is extended far enough so that one has $AP - BQ > 2AB$, and thus one will have $AP > BQ + 2AB$. But on the contrary, the straight line AP is shorter than the angular line $ABQP$, which joins the same extremities A and P, so that one has always $AP < AB + BQ + QP$; thus the hypothesis with which one started is absurd; therefore the sum of the three angles in triangle ABC is not greater than two right angles.

PROPOSITION XX.
THEOREM.

In any triangle, the sum of the three angles is equal to two right angles.

Having already proved that the sum of the three angles of a triangle cannot exceed two right angles, it remains to prove that this same sum cannot be smaller than two right angles.

Let ABC be the proposed triangle [Fig. 1.10], and let, if possible, the sum of its angles $= 2P - Z$, where P denotes a right angle, and Z is whatever quantity by which one supposes the angle sum is less than two right angles.

Let A be the smallest of the angles in triangle ABC, on the opposite side BC make the angle $BCD = ABC$, and the angle $CBD = ACB$; the triangles BCD, ABC will be equal, by having an equal side BC adjacent to two corresponding

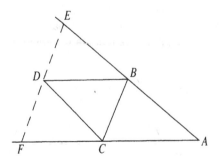

FIGURE 1.10. Legendre's refutation of HAA.

equal angles [pr. 7]. Through the point D draw any straight line EF that meets the two extended sides of angle A in E and F.*

Because the sum of the angles of each of the triangles ABC, BCD is $2P - Z$, and that of each of the triangles EBD, DCF cannot exceed $2P$ [pr. 19], it follows that the sum of the angles of the four triangles ABC, BCD, EBD, DCF does not exceed $4P - 2Z + 4P$, or $8P - 2Z$. If from this sum one subtracts those of the angles at B, C, D, which is $6P$, because the sum of the angles formed at each of the points B, C, D is $2P$ [pr. 2, cor. 3], the remainder will equal the sum of the angles of triangle AEF; therefore the sum of the angles of triangle AEF does not exceed $8P - 2Z - 6P$, or $2P - 2Z$. Thus while it is necessary to add Z to the sum of the angles in triangle ABC in order to make two right angles, it is necessary to add at least $2Z$ to the sum of the angles of triangle AEF in order to likewise make two right angles.

By means of the triangle AEF one constructs in like manner a third triangle, such that it will be necessary to add at least $4Z$ to the sum of its three angles in order for the whole to equal two right angles; and by means of the third one constructs similarly a fourth, to which it will be necessary to add at least $8Z$ to the sum of its angles, in order for the whole to equal two right angles, and so forth.

Now, no matter how small Z is in relation to the right angle P, the sequence Z, $2Z$, $4Z$, $8Z$, etc., in which the terms increase by a doubling ratio, leads before long to a term equal to $2P$ or greater than $2P$. One will consequently then reach a triangle to which it will be necessary to add to the sum of its angles a quantity equal to or greater than $2P$, in order for the total sum to be just $2P$. This consequence is obviously absurd; therefore the hypothesis with which one started cannot manage to continue to exist, that is, it is impossible that the sum of the angles of triangle ABC is less than two right angles; it cannot be greater by virtue of the preceding proposition; thus it is equal to two right angles.

Legendre's leisurely exposition of his proof, a model of simplicity, is very attractive and believable, but we hope the reader has discovered the flaw. If not, try again now before reading further (Exercise 1.13).

It is telling in and of itself that the flaw is precisely at the spot where Legendre himself felt the need to include an encouraging footnote to help convince the reader. He hoped that by pointing out that angle A is smaller than two-thirds of a right angle, the reader would have no quarrel with Legendre drawing a line through the point D, interior to the angle, which meets both sides (extended as needed) of the angle. But in fact, the ability to do this is not something Legendre would be able to prove for any angle, no matter how small, without assuming the parallel postulate in advance. Thus his proof of the parallel postulate, via the angle sum theorem, became a circular argument.

---

*One supposes that A is the smallest of the angles in triangle ABC, and that consequently it is smaller or not larger than two-thirds of a right angle, so as to make more perceptible the possibility that a straight line drawn through the point D meets both the extended sides AB, AC.

Legendre believed that inside any angle less than two right angles he could draw a line through any point that meets both sides of the angle; we will call this *Legendre's postulate*. The reader will be able to prove this assumption from the parallel postulate, and also vice versa (Exercises 1.14 and 1.15), and thus Legendre's postulate is equivalent to assuming the parallel postulate. Having grown up so steeped in the worldview of Euclidean geometry, it is hard for any of us at first, as it was impossible for Legendre, to imagine that his postulate might fail, i.e., that it could be possible that no line from a given interior point D would meet both sides of an angle. Legendre suffered, as much as the rest of us sometimes do, from too strong a belief in the thing he was trying to prove, based on his strong belief in absolute space. He was blind to the possibility of a geometric world in which his postulate, and the parallel postulate, might be replaced by a different type of geometry. Much later, after Legendre had abandoned this proof in later editions of his textbook, he wrote about it, specifically referring to the use of his postulate:

> We have thus approached much closer to our goal, but we have not entirely attained it, because our proof depended on a *postulate* that could by all means be denied....

> In examining these matters with more attention, we have been left persuaded that to completely demonstrate our *postulate* it would be necessary to deduce from the definition of a straight line a characteristic property of this line that excludes all resemblance to the form of a hyperbola contained between its two asymptotes [108, 12th ed. (1823), Note II, pp. 278–79].

Here he recognizes that to obtain his postulate he would need to know that lines cannot approach each other asymptotically, and he believes that somehow this will be obtained from the very definition of a straight line, not realizing that what constitutes a straight line must remain precisely one of the undefined terms in geometry. It is rather the fundamental assumed properties of lines that determine their nature and the roles they play. Legendre's postulate, like the parallel postulate, is such a property, and must be part of the unproven assumed nature of lines, as we will see shortly in the work of others. Legendre nonetheless proceeds to try to deduce his postulate from a more basic understanding of the nature of a line, and, in defense of his postulate, is finally reduced to saying:

> It is repugnant to the nature of the straight line that such a line indefinitely extended, could be contained in an angle [108, 12th ed. (1823), Note II, p. 279].

He continues with an explanation of why such behavior would be contrary to the nature of a straight line, because of how it would result in what he considers to be unequal partitioning of the plane, and finally he claims that his argument has fully established his *postulate*, and thus the parallel postulate. We in hindsight, however, recognize this as the final "end of the line," so to speak, for a failed approach.

Legendre's work could have opened up vistas of what a geometry without the parallel postulate would be like. But he, like all before him, was fixated on establishing his belief in it as fact, attempting time and time again to perfect proofs of it, right up until his death at age 80 in 1833. At the very same time, several much younger mathematicians of the first three decades of the nineteenth century were opening wide the doors to a non-Euclidean worldview, and it is there our next source will take us.

Exercise 1.13: Before reading our commentary after Legendre's results, find and discuss the flaw in his proof of the parallel postulate. Hint: Our preceding discussion should tell you which proposition the flaw is in.

Exercise 1.14: From Euclid's parallel postulate, prove Legendre's postulate used in his proof.

Exercise 1.15: From Legendre's postulate used in his proof, prove Euclid's parallel postulate.

1.4 Lobachevskian Geometry

It is tempting to speculate that it was precisely *because* of his isolation in a remote town in Siberia that Nikolai Ivanovich Lobachevsky was able to conceive of the possibility of a non-Euclidean geometry. Aside from a considerable portion of genius, such a feat requires a certain innocence of thinking and lack of prejudice, which might have been more abundant away from the great centers of mathematics. Lobachevsky spent his whole academic career at Kazan University, first as a student, then as professor as well as president. One of his teachers there was J. C. Bartels, who had once been Gauss's high-school teacher. One might speculate that he had interested Lobachevsky in Gauss's early investigations on the parallel postulate, but there is no evidence for it. While Lobachevsky worked on other subjects, his geometric work is his most outstanding accomplishment.

After his first article on the subject, *On the Principles of Geometry*, was published in the local *Kazan Messenger* in 1829, the reaction was devastating. In a review for the Petersburg Academy of Sciences, the eminent Mikhail Ostrogradskii concluded that the paper was not worth the attention of the Academy [144, p. 209]. When in 1831 the great scientist Alexander von Humboldt visited Kazan on his journey through Siberia, there is no record that he met with the man who was shaking the very foundations of mathematics at the time [121, pp. 141 f.]. Translations of his publications about his new geometry in German and French journals did not receive serious attention until after his death. The sole exception was Gauss, who immediately recognized a kindred spirit, and who studied Russian for the express purpose of reading Lobachevsky's works in the original. Alas, as we discussed in the Introduction, Gauss had his own reasons for keeping his admiration private.

A contributing factor to the less than enthusiastic reception of his work was the considerable lack of clarity in Lobachevsky's earlier publications. His best exposition of Lobachevskian, or hyperbolic, geometry was given in his little 1840 book *Geometrische Untersuchungen zur Theorie der Parallellinien* (Geometrical Re-

PHOTO 1.6. Lobachevsky.

searches on the Theory of Parallels), published in German in Berlin. Following are excerpts from his work, dealing with plane geometry. Beginning with a definition of parallels quite similar to that of Gauss mentioned in the Introduction, he proceeds to draw conclusions about the possible relationships of parallel lines (Exercise 1.16). Much of the work focuses on spatial geometry, and on the trigonometry of hyperbolic geometry. His book is eminently accessible, and the reader is greatly encouraged to study the rest of Lobachevsky's account. The English translation given below is by G. Halsted, reprinted in [18].

Lobachevsky, from
Geometrical Researches on the Theory of Parallels

In geometry I find certain imperfections which I hold to be the reason why this science, apart from transition into analytics, can as yet make no advance from that state in which it has come to us from Euclid.

As belonging to these imperfections, I consider the obscurity in the fundamental concepts of the geometrical magnitudes and in the manner and method of representing the measuring of these magnitudes, and finally the momentous gap in the theory of parallels, to fill which all efforts of mathematicians have been so far in vain.

For this theory Legendre's endeavors have done nothing, since he was forced to leave the only rigid way to turn into a side path and take refuge in auxiliary theorems which he illogically strove to exhibit as necessary axioms. My first essay on the foundations of geometry I published in the Kazan *Messenger* for the year 1829. In the hope of having satisfied all requirements, I undertook hereupon a treatment of the whole of this science, and published my work in separate parts in the *"Gelehrten Schriften der Universitæt Kasan"* for the years 1836, 1837, 1838, under the title "New Elements of Geometry, with a complete Theory of Parallels." The extent of this work perhaps hindered my countrymen from following such a subject, which since Legendre had lost its interest. Yet I am of the opinion that the Theory of Parallels should not lose its claim to the attention of geometers, and therefore I aim to give here the substance of my investigations, remarking beforehand that contrary to the opinion of Legendre, all other imperfections—for example, the definition of a straight line—show themselves foreign here and without any real influence on the theory of parallels.

In order not to fatigue my reader with the multitude of those theorems whose proofs present no difficulties, I prefix here only those of which a knowledge is necessary for what follows.

1. A straight line fits upon itself in all its positions. By this I mean that during the revolution of the surface containing it the straight line does not change its place if it goes through two unmoving points in the surface: (*i.e.*, if we turn the surface containing it about two points of the line, the line does not move.)

2. Two straight lines can not intersect in two points.

3. A straight line sufficiently produced both ways must go out beyond all bounds, and in such way cuts a bounded plain into two parts.

4. Two straight lines perpendicular to a third never intersect, how far soever they be produced.

5. A straight line always cuts another in going from one side of it over to the other side: (*i.e.*, one straight line must cut another if it has points on both sides of it.)

6. Vertical angles, where the sides of one are productions of the sides of the other, are equal. This holds of plane rectilineal angles among themselves, as also of plane surface angles: (*i.e.*, dihedral angles.)

7. Two straight lines can not intersect if a third cuts them at the same angle.

8. In a rectilineal triangle equal sides lie opposite equal angles, and inversely.

9. In a rectilineal triangle, a greater side lies opposite a greater angle. In a right-angled triangle the hypothenuse is greater than either of the other sides, and the two angles adjacent to it are acute.

10. Rectilineal triangles are congruent if they have a side and two angles equal, or two sides and the included angle equal, or two sides and the angle opposite the greater equal, or three sides equal.

. . .

From here follow the other theorems with their explanations and proofs.

16. All straight lines which in a plane go out from a point can, with reference to a given straight line in the same plane, be divided into two classes—into *cutting* and *not-cutting*.

The *boundary lines* of the one and the other class of those lines will be called *parallel to the given line*.

From the point A (Fig. 1.11) let fall upon the line BC the perpendicular AD, to which again draw the perpendicular AE.

In the right angle EAD either will all straight lines which go out from the point A meet the line DC, as for example AF, or some of them, like the perpendicular AE, will not meet the line DC. In the uncertainty whether the perpendicular AE is the only line which does not meet DC, we will assume it may be possible that there are still other lines, for example AG, which do not cut DC, how far soever they may be prolonged. In passing over from the cutting lines, as AF, to the non-cutting lines, as AG, we must come upon a line AH, parallel to DC, a boundary line, upon one side of which all lines AG are such as do not meet the line DC, while upon the other side every straight line AF cuts the line DC.

The angle HAD between the parallel HA and the perpendicular AD is called the parallel angle (angle of parallelism), which we will here designate by $\Pi(p)$ for $AD = p$.

If $\Pi(p)$ is a right angle, so will the prolongation AE' of the perpendicular AE likewise be parallel to the prolongation DB of the line DC, in addition to which we remark that in regard to the four right angles, which are made at the point A by the perpendiculars AE and AD, and their prolongations AE' and AD', every straight line which goes out from the point A, either itself or at least its prolongation, lies in one of the two right angles which are turned toward BC, so that except for the parallel EE' all others, if they are sufficiently produced both ways, must intersect the line BC.

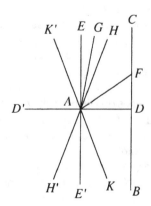

FIGURE 1.11. Lobachevsky's boundary lines.

If $\Pi(p) < \frac{1}{2}\pi$,[5] then upon the other side of AD, making the same angle $DAK = \Pi(p)$ will lie also a line AK, parallel to the prolongation DB of the line DC, so that under this assumption we must also make a distinction of *sides in parallelism*.

All remaining lines or their prolongations within the two right angles turned toward BC pertain to those that intersect, if they lie within the angle $HAK = 2\Pi(p)$ between the parallels; they pertain on the other hand to the non-intersecting AG, if they lie upon the other sides of the parallels AH and AK, in the opening of the two angles $EAH = \frac{1}{2}\pi - \Pi(p)$, $E'AK = \frac{1}{2}\pi - \Pi(p)$, between the parallels and EE' the perpendicular to AD. Upon the other side of the perpendicular EE' will in like manner the prolongations AH' and AK' of the parallels AH and AK likewise be parallel to BC; the remaining lines pertain, if in the angle $K'AH'$, to the intersecting, but if in the angles $K'AE$, $H'AE'$ to the non-intersecting.

In accordance with this, for the assumption $\Pi(p) = \frac{1}{2}\pi$ the lines can be only intersecting or parallel; but if we assume that $\Pi(p) < \frac{1}{2}\pi$, then we must allow two parallels, one on the one and one on the other side; in addition we must distinguish the remaining lines into non-intersecting and intersecting.

For both assumptions it serves as the mark of parallelism that the line becomes intersecting for the smallest deviation toward the side where lies the parallel, so that if AH is parallel to DC, every line AF cuts DC, how small soever the angle HAF may be.

17. *A straight line maintains the characteristic of parallelism at all its points.*

Given AB (Fig. 1.12) parallel to CD, to which latter AC is perpendicular.

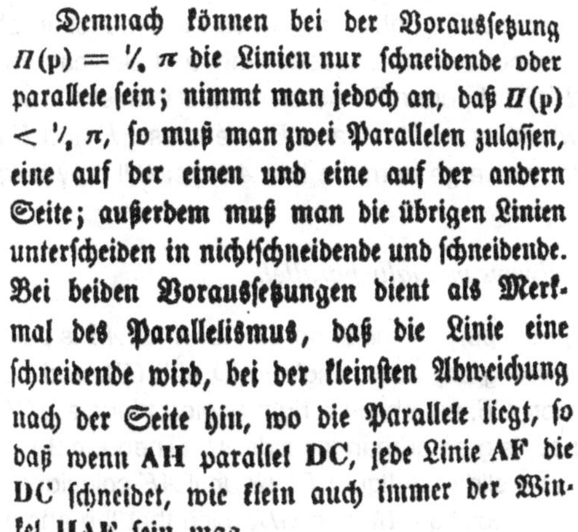

Demnach können bei der Voraussetzung
$\Pi(p) = \frac{1}{2}\pi$ die Linien nur schneidende ober
parallele sein; nimmt man jedoch an, daß $\Pi(p)$
$< \frac{1}{2}\pi$, so muß man zwei Parallelen zulassen,
eine auf der einen und eine auf der andern
Seite; außerdem muß man die übrigen Linien
unterscheiden in nichtschneidende und schneidende.
Bei beiden Voraussetzungen dient als Merk-
mal des Parallelismus, daß die Linie eine
schneidende wird, bei der kleinsten Abweichung
nach der Seite hin, wo die Parallele liegt, so
daß wenn AH parallel DC, jede Linie AF die
DC schneidet, wie klein auch immer der Win-
kel HAF sein mag.

PHOTO 1.7. Geometrische Untersuchungen ...

---

[5]Here, $\pi/2$ is Lobachevsky's measure for a right angle.

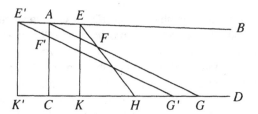

FIGURE 1.12. The characteristic of parallelism.

We will consider two points at random on the line AB and its production beyond the perpendicular.

Let the point E lie on that side of the perpendicular on which AB is looked upon as parallel to CD.

Let fall from the point E a perpendicular EK on CD and so draw EF that it falls within the angle BEK.

Connect the points A and F by a straight line, whose production then (by Theorem 16) must cut CD somewhere in G. Thus we get a triangle ACG, into which the line EF goes; now since this latter, from the construction, can not cut AC, and can not cut AG or EK a second time (Theorem 2), therefore it must meet CD somewhere at H (Theorem 3).

Now let E' be a point on the production of AB and $E'K'$ perpendicular to the production of the line CD; draw the line $E'F'$ making so small an angle $AE'F'$ that it cuts AC somewhere in F'; making the same angle with AB, draw also from A the line AF, whose production will cut CD in G (Theorem 16.)

Thus we get a triangle AGC, into which goes the production of the line $E'F'$; since now this line can not cut AC a second time, and also can not cut AG, since the angle $BAG = BE'G'$ (Theorem 7), therefore must it meet CD somewhere in G'.

Therefore, from whatever points E and E' the lines EF and $E'F'$ go out, and however little they may diverge from the line AB, yet will they always cut CD, to which AB is parallel.

18. *Two lines are always mutually parallel.*

Let AC (Fig. 1.13) be a perpendicular on CD, to which AB is parallel. If we draw from C the line CE making any acute angle ECD with CD, and let fall from A the perpendicular AF upon CE, we obtain a right-angled triangle ACF, in which AC, being the hypothenuse, is greater than the side AF (Theorem 9.)

Make $AG = AF$, and slide the figure $EFAB$ until AF coincides with AG, when AB and FE will take the position AK and GH, such that the angle $BAK = FAC$, consequently AK must cut the line DC somewhere in K (Theorem 16), thus forming a triangle AKC, on one side of which the perpendicular GH intersects the line AK in L (Theorem 3), and thus determines the distance AL of the intersection point of the lines AB and CE on the line AB from the point A.

Hence it follows that CE will always intersect AB, how small soever may be the angle ECD, consequently CD is parallel to AB (Theorem 16).

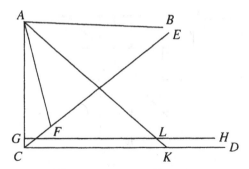

FIGURE 1.13. Mutual parallelism.

19. *In a rectilineal triangle the sum of the three angles can not be greater than two right angles.*

. . .

20. *If in any rectilineal triangle the sum of the three angles is equal to two right angles, so is this also the case for every other triangle.*

If in the rectilineal triangle ABC (Fig. 1.14) the sum of the three angles $= \pi$, then must at least two of its angles, A and C, be acute. Let fall from the vertex of the third angle B upon the opposite side AC the perpendicular p. This will cut the triangle into two right-angled triangles, in each of which the sum of the three angles must also be π, since it can not in either be greater than π, and in their combination not less than π.

So we obtain a right-angled triangle with the perpendicular sides p and q, and from this a quadrilateral whose opposite sides are equal and whose adjacent sides p and q are at right angles (Fig. 1.15).

By repetition of this quadrilateral we can make another with sides np and q, and finally a quadrilateral $ABCD$ with sides at right angles to each other, such that $AB = np$, $AD = mq$, $DC = np$, $BC = mq$, where m and n are any whole numbers. Such a quadrilateral is divided by the diagonal DB into two congruent right-angled triangles, BAD and BCD, in each of which the sum of the three angles $= \pi$.

The numbers n and m can be taken sufficiently great for the right-angled triangle ABC (Fig. 1.16) whose perpendicular sides $AB = np$, $BC = mq$, to enclose within itself another given (right angled) triangle BDE as soon as the right angles fit each other.

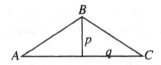

FIGURE 1.14. Angle sum π.

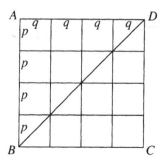

FIGURE 1.15. Quadrilateral *ABCD*.

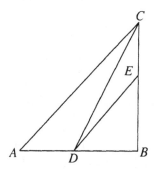

FIGURE 1.16. Enclosing triangle *BDE*.

Drawing the line DC, we obtain right-angled triangles of which every successive two have a side in common.

The triangle ABC is formed by the union of the two triangles ACD and DCB, in neither of which can the sum of the angles be greater than π; consequently it must be equal to π, in order that the sum in the compound triangle may be equal to π.

In the same way the triangle BDC consists of the two triangles DEC and DBE, consequently must in DBE the sum of the three angles be equal to π, and in general this must be true for every triangle, since each can be cut into two right-angled triangles.

From this it follows that only two hypotheses are allowable: Either is the sum of the three angles in all rectilineal triangles equal to π, or this sum is in all less than π.

21. *From a given point we can always draw a straight line that shall make with a given straight line an angle as small as we choose.*

Let fall from the given point A (Fig. 1.17) upon the given line BC the perpendicular AB; take upon BC at random the point D; draw the line AD; make $DE = AD$, and draw AE.

In the right-angled triangle ABD let the angle $ADB = \alpha$; then must in the isosceles triangle ADE the angle AED be either $\frac{1}{2}\alpha$ or less (Theorems 8 and 20).

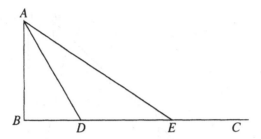

FIGURE 1.17. Small angles.

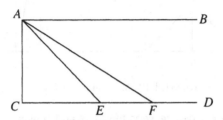

FIGURE 1.18. Parallel perpendiculars.

Continuing thus we finally attain to such an angle *AEB*, as is less than any given angle.

22. *If two perpendiculars to the same straight line are parallel to each other, then the sum of the three angles in a rectilineal triangle is equal to two right angles.*

Let the lines *AB* and *CD* (Fig. 1.18) be parallel to each other and perpendicular to *AC*.

Draw from *A* the lines *AE* and *AF* to the points *E* and *F*, which are taken on the line *CD* at any distances *FC* > *EC* from the point *C*.

Suppose in the right-angled triangle *ACE* the sum of the three angles is equal to $\pi - \alpha$, in the triangle *AEF* equal to $\pi - \beta$, then must it in triangle *ACF* equal $\pi - \alpha - \beta$, where α and β can not be negative.

Further, let the angle $BAF = a$, $AFC = b$, so is $\alpha + \beta = a - b$; now by revolving the line *AF* away from the perpendicular *AC* we can make the angle a between *AF* and the parallel *AB* as small as we choose; so also can we lessen the angle b, consequently the two angles α and β can have no other magnitude than $\alpha = 0$ and $\beta = 0$.

It follows that in all rectilineal triangles the sum of the three angles is either π and at the same time also the parallel angle $\Pi(p) = \frac{1}{2}\pi$ for every line p, or for all triangles this sum is $< \pi$ and at the same time also $\Pi(p) < \frac{1}{2}\pi$.

The first assumption serves as *foundation for the ordinary geometry and plane trigonometry.*

The second assumption can likewise be admitted without leading to any contradiction in the results, and founds a new geometric science, to which I have given the name *Imaginary Geometry*, and which I intend here to expound as far as the

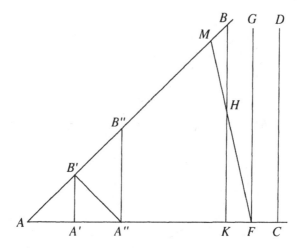

FIGURE 1.19. Angle of parallelism.

development of the equations between the sides and angles of the rectilineal and spherical triangle.

23. *For every given angle* α *there is a line p such that* $\Pi(p) = \alpha$.

Let AB and AC (Fig. 1.19) be two straight lines which at the intersection point A make the acute angle α; take at random on AB a point B'; from this point drop $B'A'$ at right angles to AC; make $A'A'' = AA'$; erect at A'' the perpendicular $A''B''$; and so continue until a perpendicular CD is attained, which no longer intersects AB. This must of necessity happen, for if in the triangle $AA'B'$ the sum of all three angles is equal to $\pi - a$, then in the triangle $AB'A''$ it equals $\pi - 2a$, in triangle $AA''B''$ less than $\pi - 2a$ (Theorem 20), and so forth, until it finally becomes negative and thereby shows the impossibility of constructing the triangle.

The perpendicular CD may be the very one nearer than which to the point A all others cut AB; at least in the passing over from those that cut to those not cutting such a perpendicular FG must exist.

Draw now from the point F the line FH, which makes with FG the acute angle HFG, on that side where lies the point A. From any point H of the line FH let fall upon AC the perpendicular HK, whose prolongation consequently must cut AB somewhere in B, and so makes a triangle AKB, into which the prolongation of the line FH enters, and therefore must meet the hypotenuse AB somewhere in M. Since the angle GFH is arbitrary and can be taken as small as we wish, therefore FG is parallel to AB and $AF = p$. (Theorems 16 and 18.)

One easily sees that with the lessening of p the angle α increases, while for $p = 0$, it approaches the value $\frac{1}{2}\pi$; with the growth of p the angle α decreases, while it continually approaches zero for $p = \infty$.

Since we are wholly at liberty to choose what angle we will understand by the symbol $\Pi(p)$ when the line p is expressed by a negative number, so we will assume

$$\Pi(p) + \Pi(-p) = \pi,$$

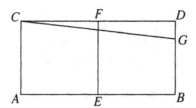

FIGURE 1.20. Approaching parallels.

an equation which shall hold for all values of p, positive as well as negative, and for $p = 0$.

24. *The farther parallel lines are prolonged on the side of their parallelism, the more they approach one another.*

If to the line AB (Fig. 1.20) two perpendiculars $AC = BD$ are erected and their end-points C and D joined by a straight line, then will the quadrilateral $CABD$ have two right angles at A and B, but two acute angles at C and D (Theorem 22) which are equal to one another, as we can easily see by thinking the quadrilateral super-imposed upon itself so that the line BD falls upon AC and AC upon BD.

Halve AB and erect at the mid-point E the line EF perpendicular to AB. This line must also be perpendicular to CD, since the quadrilaterals $CAEF$ and $FDBE$ fit one another if we so place one on the other that the line EF remains in the same position. Hence the line CD can not be parallel to AB, but the parallel to AB for the point C, namely CG, must incline toward AB (Theorem 16) and cut from the perpendicular BD a part $BG < CA$.

Since C is a random point in the line CG, it follows that CG itself nears AB the farther it is prolonged.

From here Lobachevsky proceeds to work out a wealth of trigonometric formulae, investigating the angle of parallelism $\Pi(x)$ in relation to angles in triangles. One may view this angle $\Pi(x)$ as a function from the positive real numbers to the interval $(0, \pi/2)$. Just to mention one example of his truly beautiful formulas, he shows later (p. 41) that

$$\tan \tfrac{1}{2} \Pi(x) = e^{-x}.$$

In this formula, e is an unknown constant that depends on the choice of unit for measurement in his geometry. He says (p. 33):

Since e is an unknown number only subjected to the condition $e > 1$, and further the linear unit for x may be taken at will, therefore we may, for the simplification of reckoning, so choose it that by e is to be understood the base of Napierian logarithms.

Lobachevsky also gives a precise formula for the distance between parallel lines as they are prolonged. His Theorem 33 states:

Let $AA' = BB' = x$ (Figure 1.21) be two lines parallel toward the side from A to A', which parallels serve as axes for the two boundary arcs (arcs on two boundary

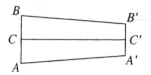

FIGURE 1.21. Asymptotic parallels.

lines) $AB = s$, $A'B' = s'$, then is

$$s' = se^{-x}$$

where e is independent of the arcs s, s' and of the straight line x, the distance of the arc s' from s.

We will not explain here exactly what these boundary arcs are, which he uses to measure distances between parallel lines, except to say that they are curves (not straight lines) that he calls "circles with infinitely great radius" (Exercises 1.19–1.20). About his formula for how parallels approach each other, he says (p. 33):

We may here remark, that $s' = 0$ for $x = \infty$, hence not only does the distance between two parallels decrease (Theorem 24), but with the prolongation of the parallels toward the side of parallelism this at last wholly vanishes. Parallel lines have therefore the character of asymptotes.

Lobachevsky's book contains many more results about two- and three-dimensional hyperbolic geometry, and we encourage the reader to study them.

A crucial omission in Lobachevsky's work was a proof that he had indeed laid the foundation for a consistent geometry, which he based on his experience with it. When such proofs were finally given by several researchers, it greatly aided the acceptance of hyperbolic geometry. In the next section we discuss one of these consistency proofs, given by the French mathematician Henri Poincaré.

Exercise 1.16: Compare and contrast the definitions of parallelism given by Gauss and Lobachevsky. Look up Bolyai's definition and compare it to the other two.

Exercise 1.17: Study Lobachevsky's proof of Theorem 19. Compare it with the two proofs of Legendre that we have read.

Exercise 1.18: Study Theorem 25 on parallel lines and its proof.

Exercise 1.19: Study Lobachevsky's development of "boundary arcs," that is, circles of infinite radius, in Theorems 29–32.

Exercise 1.20: Study the proof of Theorem 33, and how he measures the decreasing distances between parallels.

1.5   Poincaré's Euclidean Model for Non-Euclidean Geometry

Jules Henri Poincaré was born in Nancy, in the French region of Lorraine near the German border, in an upper-middle-class family. He was a brilliant student, and entered the elite Ecole Polytechnique in Paris in 1873. After graduating, he worked briefly as an engineer while finishing his doctoral thesis in mathematics, completed in 1879. He taught for a short time at the University of Caen, became a professor at the University of Paris in 1881, and taught there until his untimely death in 1912, by which time he had received innumerable prizes and honors [42].

Poincaré was well known in his era as the most gifted expositor of mathematics and science for the layperson. His essay *Mathematical Creation* remains unsurpassed to this day as an insightful and provocative description of the mental process of mathematical discovery, exploring the interplay between the conscious and subconscious mind [130, pp. 2041–2050].

Poincaré was one of the last universal mathematicians. With a grasp of the breadth of mathematics, he contributed his genius to many of its branches, and single-handedly created several new fields. Altogether he wrote nearly five hundred research papers, along with many books and lecture notes. Throughout his work, the idea of continuity was a leitmotiv. When he attacked a problem, he would investigate what happens when the conditions of the problem vary continuously, and this brought him not only new discoveries but the inauguration of entire new areas of mathematics.

PHOTO 1.8. Poincaré.

In analysis he began the theory of "automorphic functions" (certain types of these are called Fuchsian or Kleinian by him), generalizations of the trigonometric functions, and these have played a critical role throughout twentieth-century mathematics. We shall see in this section how these discoveries led him to a beautiful Euclidean model explaining Lobachevsky's non-Euclidean hyperbolic geometry, which then itself further stimulated his work on automorphic functions.

At the center of Poincaré's thought was the theory of differential equations and its applications to subjects like celestial mechanics, and he published in this area almost annually for over thirty years. A differential equation is an equation involving functions and their derivatives (see the analysis chapter), which expresses fundamental properties of the functions, often arising from physical constraints such as the motion of celestial bodies according to Newton's law of gravitation. Poincaré initiated the qualitative theory of differential equations. One of the big questions, unresolved in general to this day, is the long-term stability of something like our solar system, i.e., whether the motions under gravity of the sun and planets will cause them to remain in periodic (repeating) orbits, or orbits remaining "close" to such stable orbits, or to fly off in some way. Poincaré used his qualitative theory of differential equations to study this problem, particularly his method of varying the conditions continuously to see what happens with small changes in the initial conditions, and was able to prove that such stable solutions do exist in certain situations.

Poincaré also made great contributions to the theory of partial differential equations for functions of several variables, which has pervasive applications throughout mathematical physics. He played a critical role in the discovery of radioactivity via insights about the connection between x-rays and phosphorescence, and was involved in the invention of special relativity theory, with many physicists considering him a coinventor with Lorentz and Einstein.

In number theory and algebra Poincaré introduced important new ideas. He wrote the first paper on what we today call "algebraic geometry over the rational numbers," which addresses the problem of finding solutions to Diophantine equations. The central theme of the number theory chapter, the equation of Fermat, is an example of a Diophantine equation, and many questions raised by Poincaré remain important, and some unanswered, a century later.

Finally, as if this weren't enough, Poincaré began the mathematical subject called algebraic topology. It emerged from his interest, described above, in whether and how varying something continuously will change its qualitative features. The behavior of these qualitative features under continuous change is what we call topology, and Poincaré began to apply ideas from algebra (e.g., the groups discovered by Galois and others earlier in the nineteenth century; see the algebra chapter) to study the topology of surfaces, including in particular the phenomena he was investigating in celestial mechanics. This marriage of algebra and topology has been one of the most potent forces throughout twentieth-century mathematics.

Poincaré's role in the story of this chapter comes via a connection he discovered between Euclidean and hyperbolic geometry while studying two other branches of mathematics. Although Lobachevsky, Bolyai, and Gauss had developed the new theory of hyperbolic geometry with confidence, still mathematicians did not feel as sure of its validity as they felt of Euclid's geometry. They worried that there could be a contradiction in this new theory, since it seemed so strange and was inconsistent with Euclidean geometry. However, a few decades after the work of Lobachevsky, an extraordinary development occurred that gave his geometry equal standing with Euclid's. Mathematicians began to find "models" (realizations) for hyperbolic geometry inside Euclidean geometry. The inescapable conclusion was that it is just as mathematically correct as Euclid's. We will explain what we mean by all this in the context of the model provided by Poincaré. While his model was not the first one discovered, it is particularly beautiful, intuitively satisfying, and easy to work with in many ways. Certain aspects of Poincaré's model had been described by Riemann and Beltrami in the 1850s and 1860s, but entirely in terms of formulas using coordinates, without geometric interpretations. Further information about this and other models for non-Euclidean geometries can be found in [78, 144, 162, 164].

Poincaré was led to his model by his work on new types of functions in analysis [162, pp. 251–254]. He was working on generalizations of the trigonometric functions and their inverse functions. The reader familiar with calculus will know that the inverse trigonometric functions are integrals (antiderivatives) of expressions involving square roots of quadratic polynomials, such as

$$\int \frac{dx}{\sqrt{1 - x^2}} = \sin^{-1}(x) + C,$$

and will recall that the trigonometric functions also have periodicity properties, e.g., $\sin(x + 2\pi) = \sin(x)$. For some time mathematicians had tried to find a theory of new types of functions with analogous properties, generalizations of the inverse trigonometric functions that would solve more general integration problems involving higher roots of polynomials of any degree. The first such functions, arising early in the nineteenth century, were called elliptic functions because they came from integrals for calculating the lengths of ellipses, which involve square roots of cubic polynomials. It was discovered that the elliptic functions had double periodicity when considered as functions of a complex variable. By representing complex numbers as points in the plane, the double periodicity of the elliptic functions was exhibited geometrically as the symmetry of the periodic repetition in parallelogram tiling patterns in the Euclidean plane. For instance, Figure 1.22 illustrates the idea that an elliptic function defined on the Euclidean plane should have the same value at corresponding points in all the parallelograms, for instance at all the points marked *. By the mid–nineteenth century, generalizations of elliptic functions were being studied that provided solutions to the more general integrals and that were central to the thriving study of differential equations. However, the repetitive symmetry patterns of these new functions did not seem to arise from geometric tiling patterns in the plane, as they had for the elliptic functions. Here is where Poincaré made a breakthrough in 1881, and we can read his very own description of it. He called the new functions he was seeking to describe and understand Fuchsian functions, after I. Fuchs, whose work had inspired his own.

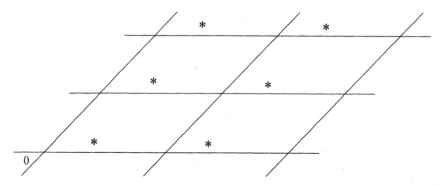

FIGURE 1.22. Double periodicity of elliptic functions.

In his famous essay *Mathematical Creation*, which still makes delightful reading today, Poincaré speaks of the psychology of mathematics in the context of his discovery relating hyperbolic geometry to Fuchsian functions:

> Just at this time I left Caen, where I was then living, to go on a geological excursion under the auspices of the school of mines. The changes of travel made me forget my mathematical work. Having reached Coutances, we entered an omnibus to go some place or other. At the moment when I put my foot on the step the idea came to me, without anything in my former thoughts seeming to have paved the way for it, that the transformations I had used to define the Fuchsian functions were identical with those of non-Euclidean geometry. I did not verify the idea; I should not have had time, as, upon taking my seat in the omnibus, I went on with a conversation already commenced, but I felt a perfect certainty. On my return to Caen, for conscience' sake I verified the result at my leisure [130, pp. 2041–2050].

In the resulting research paper *Sur les Fonctions Fuchsiennes* (On Fuchsian Functions), completed on February 14, 1881, Poincaré makes only brief mention of this important discovery:

> It was first necessary to form all the Fuchsian groups; I attained this with the assistance of non-Euclidean geometry, of which I will not speak here [133, vol. II, pp. 1–2].

Poincaré had realized that certain types of symmetries possible in the repetitive tiling patterns of the hyperbolic plane, which illustrate the "Fuchsian groups" he speaks of, would enable him to describe and study the Fuchsian functions he was seeking, particularly in light of a new model he had found for Lobachevsky's non-Euclidean geometry. Thus hyperbolic geometry served the role for his new Fuchsian functions that Euclidean geometry had served for elliptic functions.

Poincaré actually describes this new model in some detail in a slightly later paper on a completely different topic, which also turned out to have an amazing connection with hyperbolic geometry, as he says in continuing his essay on creativity:

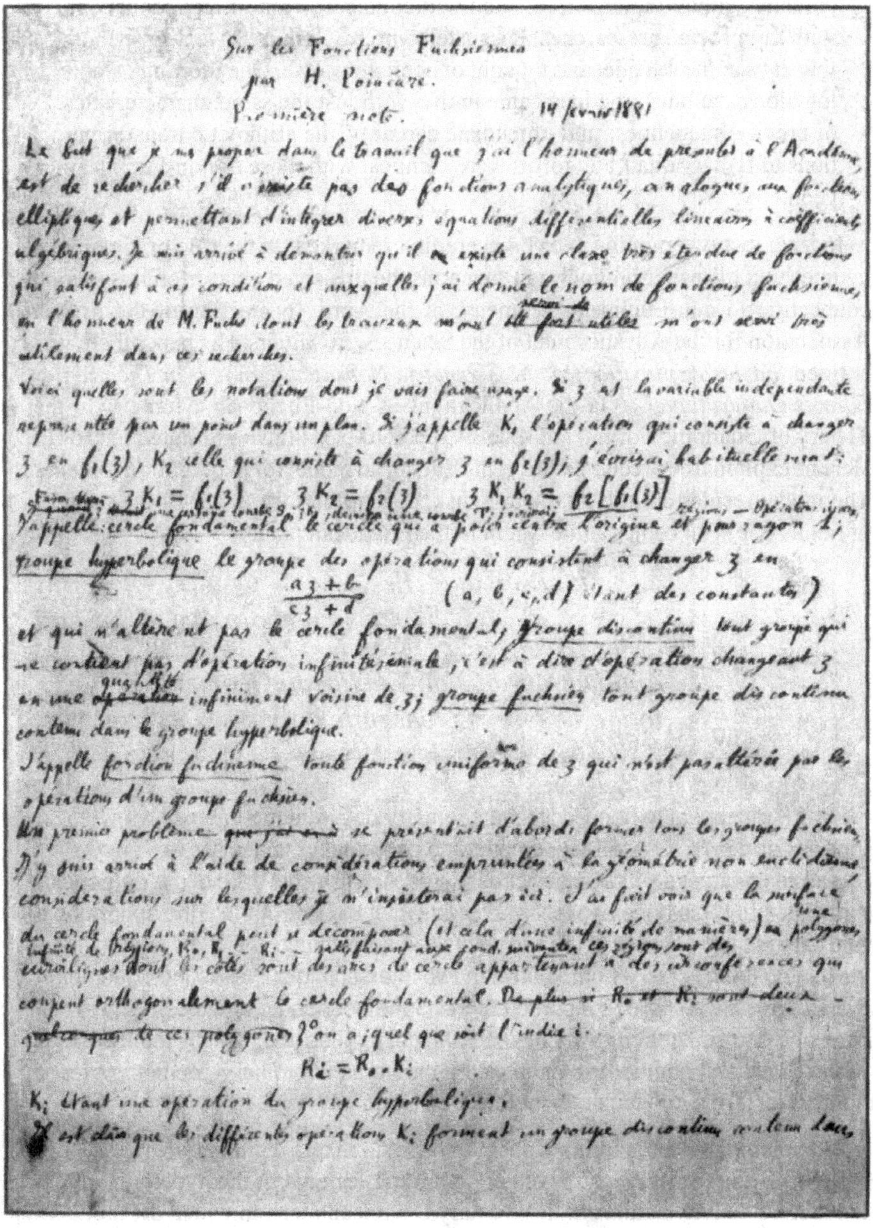

Sur les Fonctions Fuchsiennes
par H. Poincaré.
Première note.   14 février 1881

Le but que je me propose dans le travail que j'ai l'honneur de présenter à l'Académie est de rechercher s'il n'existe pas des fonctions analytiques, analogues aux fonctions elliptiques, et permettant d'intégrer diverses équations différentielles linéaires à coefficients algébriques. Je suis arrivé à démontrer qu'il en existe une classe très étendue de fonctions qui satisfont à ces conditions et auxquelles j'ai donné le nom de fonctions fuchsiennes, en l'honneur de M. Fuchs dont les travaux m'ont été fort utiles m'ont servi très utilement dans ces recherches.

Voici quelles sont les notations dont je vais faire usage. Si z est la variable indépendante représentée par un point dans un plan. Si j'appelle K_1 l'opération qui consiste à changer z en $f_1(z)$, K_2 celle qui consiste à changer z en $f_2(z)$, j'écrirai habituellement ainsi :

$$z K_1 = f_1(z) \qquad z K_2 = f_2(z) \qquad z K_1 K_2 = f_2[f_1(z)]$$

J'appelle cercle fondamental le cercle qui a pour centre l'origine et pour rayon 1 ; groupe hyperbolique le groupe des opérations qui consistent à changer z en

$$\frac{az+b}{cz+d} \qquad (a, b, c, d \text{ étant des constantes})$$

et qui n'altèrent pas le cercle fondamental ; groupe discontinu tout groupe qui ne contient pas d'opération infinitésimale, c'est à dire d'opération changeant z en une quantité infiniment voisine de z ; groupe fuchsien tout groupe discontinu contenu dans le groupe hyperbolique.

J'appelle fonction fuchsienne toute fonction uniforme de z qui n'est pas altérée par les opérations d'un groupe fuchsien.

Un premier problème qui se présentait d'abord : former tous les groupes fuchsiens. J'y suis arrivé à l'aide de considérations empruntées à la géométrie non euclidienne considérations sur lesquelles je n'insisterai pas ici. J'ai fait voir que la surface du cercle fondamental peut se décomposer (et cela d'une infinité de manières) en polygones curvilignes dont les côtés sont des arcs de cercle appartenant à des circonférences qui coupent orthogonalement le cercle fondamental. De plus si R_0 et R_i sont deux de ces polygones, l'on a, quel que soit l'indice i,

$$R_i = R_0 \cdot K_i$$

K_i étant une opération du groupe hyperbolique.

Il est clair que les différentes opérations K_i forment un groupe discontinu et leur loi…

PHOTO 1.9. *Sur les Fonctions Fuchsiennes.*

Then I turned my attention to the study of some arithmetic questions apparently without much success and without a suspicion of any connection with my previous researches. Disgusted with my failure, I went to spend a few days at the seaside, and thought of something else. One morning, walking along the bluff, the idea came to me, with just the same characteristics of brevity, suddenness and immediate certainty, the arithmetic transformations of ternary quadratic forms were identical with those of non-Euclidean geometry.

Indeed, just two months after Poincaré first remarked on the role of hyperbolic geometry in his paper on Fuchsian functions, he presented a paper on this second connection to non-Euclidean geometry, at the April 16 meeting of the French Association for the Advancement of the Sciences, in Algiers. His paper there was entitled *Sur les Applications de la Géométrie Non Euclidienne a la Théorie des Formes Quadratiques* (On the Applications of Non-Euclidean Geometry to the Theory of Quadratic Forms) [133, vol. V, pp. 270–271] (also translated in [164]). Here he explains the precise nature of his Euclidean model for hyperbolic geometry. The model resides entirely in the unit disk, constituting the inside of the circle C of radius one with center at the origin in the Euclidean plane.

Poincaré, from

On the Applications of Non-Euclidean Geometry to the Theory of Quadratic Forms

Here I will appeal to non-Euclidean geometry or pseudogeometry. I will write, in short, *ps* and *psly*, for pseudogeometry and pseudogeometrically.
I will designate, as a *ps* line, any circumference which cuts the circle C perpendicularly; as the *ps* distance between two points, half the logarithm of the anharmonic ratio of these two points with the two points of intersection of the circle C with the *ps* line joining them. The *ps* angle between two intersecting curves is their geometric angle. A *ps* polygon will be a portion of the plane bounded by *ps* lines....
. . . one finds out that *ps* distances, *ps* angles, *ps* lines, etc., satisfy the theorems of non-Euclidean geometry, that is, all the theorems of ordinary geometry, except those which are a consequence of Euclid's *postulate*.

To explain the details of his model, let us first make a dictionary (Figure 1.23) for Poincaré's terms, which translates geometric features in the hyperbolic plane into corresponding features in the ordinary Euclidean plane in which the model is embedded. Then we may use the dictionary to explore properties of non-Euclidean geometry by working simply with their Euclidean counterparts.

With this dictionary we can consider Figure 1.24, which illustrates some of the features of Poincaré's model. It shows five *ps* lines. Three of them intersect in three *ps* points, thus forming a *ps* triangle, and the other two lines do not intersect any of the other lines shown. *Ps* points, *ps* lines, and *ps* angles are extremely

Poincaré's disk model of the hyperbolic plane	Euclidean plane
a *ps* point	a Euclidean point inside the unit disk
a *ps* line	the portion inside the unit disk of any Euclidean circle meeting *C* perpendicularly, or a diameter of the circle *C*
a *ps* angle between *ps* lines	the Euclidean angle between the two Euclidean curves that are the *ps* lines
a *ps* distance between *ps* points	given by a particular formula that involves the distance of the points from the circle *C*

FIGURE 1.23. Dictionary for Poincaré's disk model.

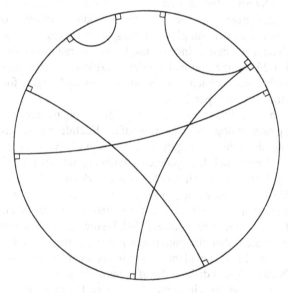

FIGURE 1.24. Poincaré's unit disk model.

easy to conceive of and work with in this model, and the reader will easily be able to explore each of the fundamental questions addressed by Euclid, Saccheri, Lambert, Legendre, and Lobachevsky.

Before suggesting how to do this, we should first note that *ps* distance is not quite as straightforward, but that many of the features we are interested in can be observed without worrying about distance, or by knowing just a little about it. Distance is given by a formula that, roughly speaking, amounts to stretching the Euclidean distance in the disk by the factor $\frac{1}{1-x^2-y^2}$ at the point with coordinates (x, y). Thus there is no stretching at the center of the disk (i.e., a stretching factor of 1), and the stretching increases as we move outwards. Halfway towards the edge of the disk along a radius, the stretching factor is $\frac{1}{1-(1/2)^2} = \frac{4}{3}$. As we move closer to the edge, the stretching increases dramatically, approaching infinite stretching as we approach the edge of the disk. The nature of the stretching in the distance formula actually causes all non-meeting *ps* lines to grow farther apart in the *ps* geometry as they approach the edge of the disk in the Euclidean plane, except for those that meet in the Euclidean sense at a point on the boundary circle *C* (such as a pair shown in Figure 1.24). These latter will actually become closer and closer as they approach *C*, but of course they do not actually meet in the *ps* world of the model: the point they appear to be approaching on the boundary *C* is not a *ps* point, since it is not inside the disk. According to the distance stretching formula, each of the *ps* lines has infinite length inside the disk. (The reader can prove this from the stretching factor formula using some calculus.)

Note that diameters of the circle *C* are also *ps* lines; it is often useful to choose one or two diameters as *ps* lines in a construction, if possible. Using the model one can now verify that all of Euclid's postulates except the parallel postulate are satisfied (Exercise 1.21). Moreover, the parallel postulate does not hold, but is replaced by Lobachevsky's alternative (Exercise 1.22). We then know that our model exhibits Lobachevsky's geometry as a completely valid system, provided that we accept the validity of Euclidean geometry, since we used it as our framework for the model.[6] Exercises 1.23–1.26 use the unit disk model to explore the questions studied by Saccheri, Legendre, and Lobachevsky about lengths and angles for hyperbolic quadrilaterals, triangles, and parallels.

Let us end with three illustrations of the repetitive tiling patterns possible in the non-Euclidean plane, analogous to the tiling of the Euclidean plane created by the collection of Euclidean lines in Figure 1.22. These will reveal some of the more curious features of non-Euclidean geometry. Poincaré himself remarked in 1882, when discussing the historical origins of his own work on Fuchsian functions and the corresponding Fuchsian groups [133, vol. II, pp. 168–169], that one of his examples of a Fuchsian group already existed in geometric form as a tiling pattern published by H. Schwarz ten years earlier [154]. Figure 1.25 comes from this paper, and we see immediately that although it is a priori a pattern of curved segments in the unit disk in the Euclidean plane, it is actually a collection of *ps* lines when viewed using Poincaré's unit disk model of hyperbolic geometry.

Whereas the Euclidean lines in Figure 1.22 divide the Euclidean plane up into infinitely many congruent parallelograms, Schwarz's *ps* lines divide the hyperbolic plane into infinitely many triangles. And although it may not seem so at first sight, in fact all the triangles are *ps* congruent (recall that distances are stretched as one moves outwards in the disk). It is not hard to see that each triangle has *ps* angles

---

[6]The great early-twentieth-century mathematician David Hilbert provided a firm foundation for both Euclidean and non-Euclidean geometry that shows that they are consistent, provided that the theory of the real numbers is consistent. On the other hand, Kurt Gödel (see the analysis chapter) proved in 1931 that no proof of the absolute consistency of either Euclidean or Lobachevskian non-Euclidean geometry is possible [164, p. 65]! We will forever live not knowing for certain that there is no internal contradiction in these geometries, or the real numbers. However, Poincaré's model does at least convince us that our non-Euclidean geometry is as free of contradictions as Euclidean geometry or the real numbers. In fact, Euclidean geometry is also contained in higher-dimensional extensions of Lobachevsky's geometry, so the different geometries are actually equally consistent [164, p. 38].

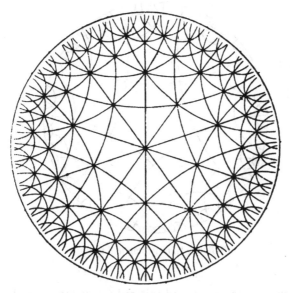

FIGURE 1.25. Schwarz's tiling.

36°, 45°, 90°, for a total angle sum of 171°. They all occur in clusters of ten fitted together around a central point to form regular right-angled pentagons. Recalling that Lambert had discovered a fixed relationship between changing *ps* side lengths and changing *ps* angles under shrinkage or expansion of triangles, so that similar *ps* triangles must actually always be *ps* congruent, we conclude that all Schwarz's triangles must be congruent, and so are the pentagons.

Removing the *ps* lines passing through the centers of the pentagons, keeping only the lines that form the edges of the pentagons, we obtain a tiling of the hyperbolic plane by regular right-angled pentagons, meeting four to a corner, shown in Figure 1.26.

Thus a hyperbolic bathroom floor can be tiled with pentagons, unlike the Euclidean floor, which can be tiled only with triangles, squares, and hexagons. A new challenge, though, emerges from the fixed relationship between angles and lengths: there is only one possible size for regular right-angled pentagons, and it might be very large or very small, creating tile-laying problems in any particular bathroom.[7] Imagine what happens if we try to simultaneously expand all the right-angled pentagonal tiles on the hyperbolic floor. Unlike Euclidean geometry, where the entire pattern can be expanded or shrunk at will with no fundamental change in its nature, our pattern will crack and break apart as it changes size,

---

[7]In fact, there are many different hyperbolic plane geometries, each with its own particular size relationship between angles and lengths, so the practical prospects of tiling the bathroom floor with right-angled pentagons will depend on which hyperbolic world one happens to live in. In fact, our own physical universe may well be one of these hyperbolic types, but with the right-angled pentagon rather large (so large that in the portion of the universe we have examined everything appears Euclidean to us for all practical purposes).

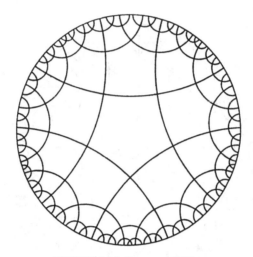

FIGURE 1.26. Pentagonal tiling.

since as the regular pentagons expand uniformly, their angles will lessen in size. Cracks will appear between the pentagons, and four will no longer fit together at a point. However, after a certain amount of expansion, their angles will have decreased to 72°, and then we can try fitting them together again, meeting five to a corner. Later on in the expansion, their angles will decrease to 60°, and we can place them together six to a corner, and so on ad infinitum. It is a remarkable fact that for any of these possibilities they can be fitted together without gaps or overlaps to create perfect tilings of the plane. We see that the hyperbolic designer has a wealth of new aesthetic tiling options that the Euclidean designer lacks, provided that the relationship between hyperbolic angle and distance is such that the scale of each particular pattern suits the designer's constraints, since each pattern comes in only one size! By enlarging the size of each item, new patterns arise, but on the other hand, shrinking the size of the items makes the possibilities disappear as the situation becomes closer and closer to the Euclidean one.

Tilings of the Poincaré model of the non-Euclidean plane have served as great inspiration in art, particularly in the work of twentieth-century Dutch graphic artist Maurits C. Escher. In Figure 1.27, his pattern of angels and devils is overlaid on a tiling with right-angled non-Euclidean hexagons, meeting four to a corner, completely analogous to the pentagonal tiling. If we divide the hexagonal tiling up into a triangular tiling analogous to that of Schwarz, then each pair of matching mutually reflecting triangles represents one of Escher's bilaterally symmetrical angels or devils (Exercise 1.27). The reader may explore all these phenomena further in [78, 162, 164].

The viewpoint opened by models like Poincaré's has greatly expanded our perception of what geometry is, and there are many other completely different types of geometries known and studied today. Moreover, their connection to other branches

FIGURE 1.27. Escher's tiling.

of mathematics and to applications is manifold, and research in various geometries is a thriving part of the mathematical enterprise today.

Exercise 1.21: Show that all of Euclid's postulates except the parallel postulate are satisfied in Poincaré's unit disk model.

Exercise 1.22: Does Euclid's parallel postulate hold in Poincaré's unit disk model? Does Lobachevsky's alternative hold?

Exercise 1.23: Which of Saccheri's three hypotheses about angles in Saccheri quadrilaterals holds in Poincaré's unit disk model?

Exercise 1.24: In Poincaré's unit disk model, how is the angle sum of a triangle related to its size? How large or small can the angle sum be? Are there similar triangles?

Exercise 1.25: In Poincaré's unit disk model, is the angle sum of a triangle always equal to, greater than, or less than two right angles, as Legendre studied? Discuss Legendre's hypothesis.

Exercise 1.26: Describe the behavior of Lobachevsky's angle of parallelism for various *ps* lines and points. How big or small can it be? Describe which *ps* lines are parallel according to Lobachevsky's definition.

Exercise 1.27: Find and describe some tilings of the hyperbolic plane in the work of M.C. Escher or other artists.

CHAPTER 2

Set Theory: Taming the Infinite

2.1 Introduction

"I see it, but I don't believe it!" This disbelief of Georg Cantor in his own creations exemplifies the great skepticism that his work on infinite sets inspired in the mathematical community of the late nineteenth century. With his discoveries he single-handedly set in motion a tremendous mathematical earthquake that shook the whole discipline to its core, enriched it immeasurably, and transformed it forever. Besides disbelief, Cantor encountered fierce opposition among a considerable number of his peers, who rejected his discoveries about infinite sets on philosophical as well as mathematical grounds.

Beginning with Aristotle (384–322 B.C.E.), two thousand years of Western doctrine had decreed that actually existing collections of infinitely many objects of any kind were not to be part of our reasoning in philosophy and mathematics, since they would lead directly into a quagmire of logical contradictions and absurd conclusions. Aristotle's thinking on the infinite was in part inspired by the paradoxes of Zeno of Elea during the fifth century B.C.E. The most famous of these asserts that Achilles, the fastest runner in ancient Greece, would be unable to surpass a much slower runner, provided that the slower runner got a bit of a head start. Namely, Achilles would then first have to cover the distance between the starting positions, during which time the slower runner could advance a certain distance. Then Achilles would have to cover that distance, while the slower runner would again advance, and so on. Even though the distances would be getting very small, there are infinitely many of them, so that it would take Achilles infinitely long to cover all of them [4, p. 179]. Aristotle deals with this and Zeno's other paradoxes (Exercise 2.1) at great length, concluding that the way to resolve them is to deny the possibility of collecting infinitely many objects into a complete and actually existing whole. The only allowable concept is the so-called potential infinite. While it is inadmissible to consider the complete collection of all natural numbers, it is allowed to consider a finite collection of such numbers that can be enlarged as much as one wishes. As an analogy, when we study a function $f(x)$ of a real variable x, then we may be interested in the behavior of $f(x)$ as x becomes arbitrarily large. But we are not allowed simply to replace x by ∞ to find out, because we might get a nonsensical result.

A paradox of a more overtly mathematical nature, given by Galileo Galilei (1564–1642) in the seventeenth century, shows that there are just as many perfect squares as there are natural numbers, by pairing off each number with its square:

$$
\begin{array}{ccccc}
1 & 2 & 3 & 4 & \cdots \\
\updownarrow & \updownarrow & \updownarrow & \updownarrow & \updownarrow \\
1 & 4 & 9 & 16 & \cdots
\end{array}
$$

But on the other hand, since not all natural numbers are perfect squares, there are clearly more natural numbers than perfect squares. Galileo even observes that the perfect squares become ever sparser as one progresses through the natural numbers, making the pairing above even more paradoxical. He concludes that "the attributes 'larger,' 'smaller,' and 'equal' have no place either in comparing infinite quantities with each other or comparing infinite with finite quantities" [65, p. 33]. A comprehensive account of the philosophical struggle with the concept of the infinite from early Greek thought to the present can be found in [123] (see also [145]).

It was left to a theologian from Prague in the early nineteenth century to make a systematic study of mathematical paradoxes involving infinity. After studying philosophy, physics, and mathematics at the University of Prague, Bernard Bolzano (1781–1848) decided that his calling was to be a theologian, even though he had been offered a chair in mathematics. As a professor of theology at the University of Prague, beginning in 1805, Bolzano nonetheless spent part of his time pursuing mathematical research. After being dismissed from his position for expressing allegedly heretical opinions in his sermons, and being prohibited from ever teaching or publishing again, he used his enforced leisure to work almost exclusively on mathematics. His philosophical interests had always drawn him to questions about the foundations of mathematics, its definitions, methods of proof, and the nature of its concepts. Thus, he naturally was led to the philosophy of the infinite. Not only did he conclude that mathematics was well equipped to deal with infinite sets in a systematic manner, free of contradictions, he even went as far as arguing that mathematics was the proper realm in which to discuss and resolve *all* paradoxes involving the infinite.[1]

---

[1] In India, on the other hand, the Jaina mathematics of the middle of the first millenium B.C.E. not only entertained the idea that many different kinds of infinity could exist, but actually developed the beginnings of a system of infinite sets and infinite numbers [91, pp. 18, 218–219, 249–253].

Bolzano made it clear that he considered the property of a *one-to-one corre-spondence* between a set and a proper subset of itself, as in Galileo's example above, fundamental to the nature of infinite sets, to be exploited in a mathematical investigation, rather than to be used as justification to avoid all discussion of the matter. By a one-to-one correspondence between two sets we mean a pairing of the elements of the sets in such a way that each element of one gets paired with a unique element of the other, without any elements in either set being left over (Exercises 2.2–2.3). The first text in this chapter is a collection of excerpts from Bolzano's book *Paradoxien des Unendlichen* (Paradoxes of the Infinite) [16], published after his death.

Bolzano worked for the most part in isolation, with little contact to the rest of the mathematical world. *Paradoxes of the Infinite* was published in 1851 in Germany, thanks to the efforts of Bolzano's friend F. Prihonsky. Unfortunately, this work, as well as his other manuscripts in analysis and geometry, failed to draw the attention of the mathematical centers in the rest of Europe until quite some time later. When the dramatic events in the theory of infinite sets started to unfold with the work of Georg Cantor (1845–1918) more than twenty years later, the motivation came from the foundations of analysis (see the analysis chapter), one of the major subjects that occupied the mathematical world during much of the nineteenth century, rather than Bolzano's pioneering insights. Since Cantor's work received a fair amount of criticism from theologians, however, the views of the theologian Bolzano were of great importance to him.

Cantor began his mathematical career in number theory, but soon became interested in analysis. During his student days in Berlin, one of his professors was the great Karl Weierstrass (1815–1897), a major figure in the theory of functions during the second half of the nineteenth century, and Cantor began working in this area. An important problem he turned to was the question whether it was possible to represent a function of a real variable by two different so-called trigonometric series, that is, certain infinite series whose terms involve trigonometric functions. Given $f(x)$, can one find real numbers a_n, b_n for $n = 0, 1, 2, \ldots$ such that

$$f(x) = \sum_{n=0}^{\infty} a_n \sin(nx) + b_n \cos(nx)?$$

Furthermore, for what values of x is this equation valid, and what conditions on $f(x)$ and the series are necessary? (See the appendix in the analysis chapter for a brief review of infinite series.) Cantor set out to show that whenever such a representation existed it was unique, under quite general conditions. In 1870, he succeeded in showing that such uniqueness followed if the series was convergent for every x with limit $f(x)$. Not satisfied with this result, he worked out several extensions during the following year. After showing first that the result held true if the hypothesis was true for all but finitely many values of x, he pushed on to show that certain infinite sets $x_1, x_2, x_3, \ldots$ of exceptional values were allowed. (For details on Cantor's work on trigonometric series see [36, Ch. 2].)

As Cantor was trying to generalize his result yet further, it became clear to him that the sorts of questions he needed to answer about infinite sets of points on the real number line required a much deeper understanding of the nature of the real numbers than was possible with the essentially geometric concept of the number line. What was needed was an "arithmetization" of the real number concept. Besides Cantor, several other mathematicians had recognized the same need and offered their own constructions of the real numbers, depending on what they perceived to be the essential property of the number line (see the analysis chapter). Richard Dedekind (1831–1916) based his construction on the insight that each real number is in some sense determined by all the rational numbers to the left and right of it; it is a "Dedekind cut," in modern terminology. Thus, the real number then could be *defined* as that pair of sets of rational numbers [38]. Dedekind succeeded in defining the arithmetic operations of addition and multiplication on such pairs and proving that they behaved just as one would expect from actual real numbers. Cantor based his definition on the idea that each real number could be defined by a sequence of rational numbers converging to it, and proposed a definition based on such sequences [28, pp. 92–102]. However, neither Dedekind nor Cantor succeeded in proving that their "real numbers" were in fact just like the geometric number line, in the sense that each point on the number line corresponded to a "number" in their system and vice versa. It is considered an *axiom*, a basic truth accepted without proof.

Equipped with his new definition, Cantor went back to study the nature of the exceptional sets of values in his uniqueness theorem, which seemed to be related to the structure of the real numbers themselves. Soon he was on his way to discoveries that would forever change his life and the nature of mathematics. In order to compare different kinds of infinite sets of real numbers, Cantor used the notion of one-to-one correspondence, like Bolzano before him. Two sets X and Y are considered to have the same *power*, in Cantor's terminology, or cardinality, if there is a one-to-one correspondence between X and Y. For instance, the set of natural numbers and the set of perfect squares considered above have the same power, since one has the one-to-one correspondence $n \leftrightarrow n^2$. The first important result Cantor stumbled across was a rather amazing fact about the power of the continuum, as the set of real numbers was called. This happened in a roundabout way, whose description requires a bit of terminology. (We follow [36, Ch. 3].)

A real number r is called *algebraic* if it satisfies an equation of the form

$$a_n x^n + a_{n-1} x^{n-1} + \cdots + a_1 x + a_0 = 0,$$

where $a_n, \ldots, a_0$ are integers. Otherwise, r is called *transcendental*. For instance, $\sqrt[3]{2}$ is algebraic, since it satisfies the equation $x^3 - 2 = 0$. Cantor had found a proof that the set of algebraic numbers had the same power as the natural numbers, that is, it was *countable*. Likewise, he had shown that the set of all rational numbers was countable as well, as we shall see below. Now he was looking for an application of this result. A good candidate was a theorem proven by the French mathematician Joseph Liouville (1809–1882) in 1851, to the effect that each interval of real numbers contained infinitely many transcendental numbers, and he explained how to produce examples. (See [112, Ch. 12] for details on Liouville's work on transcendental numbers.) Cantor gave a new proof of this result by showing that every interval of real numbers had strictly *larger* cardinality than the natural numbers, that is, it was *uncountable*. Hence, every interval had to contain infinitely many numbers that were not algebraic. In fact, virtually any number that one might randomly pick out of an interval was bound to be transcendental. When Cantor published his result in 1874 [28, pp. 115–118], it had just been proven the year before by Charles Hermite that the Euler constant e is transcendental (see [114]). (It was not until 1882 that F. Lindemann proved that π is also transcendental [10].)

Very quickly Cantor realized that the significance of his theorem went far beyond a new proof for Liouville's theorem. He had hit upon an essential characteristic of the *continuous* set of real numbers versus the *discrete* set of integers. More startling results were waiting to be discovered. To his own disbelief, he succeeded in showing that it was possible to put the points of the plane into one-to-one correspondence with the points on the line, and, more generally, the points of a space of arbitrarily high dimension could be brought into such a one-to-one correspondence with the line. This seemed to be a direct contradiction to the notion of dimension, which was thought to be a unique characteristic of a space, not to mention the fact that it went against all common sense. (The notion of dimension was later shown to be indeed characteristic of the space, because it was impossible to find a one-to-one correspondence between sets of different dimensions that was *continuous* [118].) It was this result that prompted Cantor's exclamation, "I see it but I don't believe it!" in a letter to his colleague and friend Dedekind [27]. (There is an interesting story attached to the Cantor–Dedekind correspondence. See [76].) At the end of the 1878 paper [28, pp. 119–133] in which he published these results, Cantor concludes that in order to study general infinite sets in higher-dimensional spaces, it is therefore sufficient to study *linear* sets, that is, infinite subsets of the real line. He then divides these subsets into equivalence classes, with two subsets belonging to the same class if they have the same cardinality. Thus, one such class contains the subset of natural numbers, as well as the integers (Exercise 2.2), the rational numbers, and the algebraic numbers, and all other countable subsets of the real line. There is a second class containing, among others, each interval on the real line, as well as the real numbers themselves. Cantor now makes the following rather amazing pronouncement:

> Through a method of induction, which we shall not describe here, one is led to the theorem that the number of classes resulting from this division of linear sets is finite, namely equal to *two*.

> Thus, the linear sets would consist of two classes, of which the first contains all sets that can be given the form of a function ν, where ν runs through all positive integers, while the second class contains all those sets that can be given the form of a function x, where x can assume all values ≥ 0 and ≤ 1. According to these two classes, there are only two possible powers of linear

sets. We postpone a more detailed investigation of this question until a later occasion [28, p. 132 f.].

That is, Cantor believed that any infinite subset of the continuum, as the real line was called, is either countable or is in one-to-one correspondence with the points of the closed interval [0, 1]. The thread that runs through the rest of our story traces the fate of Cantor's unsubstantiated claim, which will lead us right up to the present. In a tremendous eruption of creativity, Cantor now follows up with a series of six papers on the theory of infinite linear sets, published between 1878 and 1884 [28, pp. 139–246]. These papers represent the pinnacle of his mathematical achievements, and he will not produce anything comparable ever again. He begins the fifth of these papers, published in 1883, as follows [28, p. 165]:

> The presentation of my investigations to date in the theory of sets has reached a point where its progress depends on an extension of the concept of whole number beyond its present limits. As far as I know, this extension points in a direction that has not yet been investigated by anybody.

> My dependence on this extension of the number concept is so great that without it I would hardly be able to make the least step forward in the theory of sets. This circumstance may serve as an explanation, or as an excuse, if necessary, why I am introducing apparently foreign ideas into my investigations. Namely, I am concerned with an extension, respectively continuation, of the sequence of whole numbers beyond the infinite. As daring as this may seem, I can proclaim not only the hope, but the firm conviction, that this extension will, in time, appear rather simple, appropriate, and natural. I am quite aware that with this undertaking I am in opposition to widely held views about the mathematical infinite and frequently voiced beliefs about the essence of numbers.

As mentioned earlier, the key ingredient in the construction of his new theory of "transfinite numbers" is the tool of one-to-one correspondence and the resulting equivalence relation (Exercise 2.4). To understand the thought behind Cantor's transfinite numbers, we need to think first about finite numbers. The number 5, for instance, can be thought of as the result of abstracting from the elements of the set $\{a, b, c, d, e\}$ by ignoring their particular identity and the order in which they are given. Similarly, Cantor defines infinite "cardinal numbers" as abstractions of infinite sets, by ignoring the particular nature and order of their elements, that is, by retaining only an unordered collection of "placeholders," which are distinct, but otherwise indistinguishable. Thus, given an infinite set, such as the set of natural numbers, the associated cardinal number can be thought of as the equivalence class, under the above equivalence relation, of all sets that are in one-to-one correspondence with the set of natural numbers, that is, the collection of all those sets that are countable. Cantor called this "number" $\aleph_0$ (pronounced aleph zero). (The symbol $\aleph$ is the first letter of the Hebrew alphabet. The reason for the subscript will become clear later on.) Another way to think of $\aleph_0$ is as the collection of properties that the set of natural numbers shares with every other countable set, and only with those. Since he had shown in his 1874 paper that the set of real numbers is uncountable, there is at least one more cardinal number different from $\aleph_0$. Since the natural numbers are a proper subset of the real numbers, one would like to say that this second number is larger than $\aleph_0$.

And indeed, Cantor defines an order relation among cardinal numbers as follows. If M and N are two cardinal numbers, then we say that $M < N$ if there is a one-to-one correspondence between M and a proper subset of N but there is no subset of M that is in one-to-one correspondence with N. In trying to understand the nature of this relation, an obvious question presents itself. Are every two cardinal numbers comparable? That is, given cardinal numbers M and N, is it always true that either $N < M$, $M < N$, or $M = N$? The affirmative statement is known as the *trichotomy principle*. Or are there two sets such that neither one is equivalent to a subset of the other? While Cantor could show that if N was equivalent to a subset of M and M was equivalent to a subset of N, then M and N had to be equivalent themselves, it was going to be a long time before the full trichotomy principle could be established. Since we would expect this principle to hold for a number system, we should actually not have been so hasty to bestow this status on the collection of cardinalities.

But before expending any effort on such questions, one should make sure that there are in fact more than the two infinite cardinal numbers that we know about so far, since otherwise the question is already answered. Cantor found an absolutely ingenious and elementary proof that the power set (the set of all subsets) of any set has a cardinal number strictly larger than the cardinal number of the given set. In particular, the set of real numbers has the cardinality of the power set of the natural numbers (Exercise 2.5). With one stroke of the pen, he showed that there was in fact a whole universe of cardinal numbers out there, waiting to be explored. For these new creations to deserve being called numbers, one should, of course, be able to do arithmetic with them, that is, add them and multiply them. It turned out that union of sets made a good candidate for "addition," with the Cartesian product of sets as "multiplication." One can see that with these definitions, the usual rules of arithmetic hold, such as the commutative, associative, and distributive laws. In addition, when applied to finite cardinal numbers, these two operations reduce to ordinary addition and multiplication. Naturally, one would not expect these new numbers to behave like the finite cardinal numbers in all respects. For instance, the easily verified identity $\aleph_0 + 1 = \aleph_0$ makes for decidedly different calculations. Certain cardinal numbers simply obliterate others under addition. Similar peculiarities happen with multiplication. In fact, we already observed a rather striking one. If we call the cardinal number of the continuum $\mathfrak{c}$, then Cantor's result that the plane has the same cardinality as the line says precisely that $\mathfrak{c} \cdot \mathfrak{c} = \mathfrak{c}$, since the plane can be thought of as the Cartesian product of the line with itself.

One more arithmetic operation is needed, that of exponentiation. Define M^N to be the cardinal number of the set of all functions from N to M. As an example, let

$N = 2$, and represent 2 by the set $\{0, 1\}$. A function

$$f : N \longrightarrow \{0, 1\}$$

assigns to each element of N either 0 or 1, and defines in this way a subset of N, namely that of all elements that are assigned 1. For instance, the function

$$f : \{a, b, c\} \longrightarrow \{0, 1\},$$

which assigns the value 0 to a and 1 to b and c, corresponds to the subset $\{b, c\}$. Two distinct functions correspond to different subsets, and each subset corresponds to a function. We therefore obtain a one-to-one correspondence between the set of functions from N to $\{0, 1\}$ and the power set of N. Hence, 2^N is just the cardinal number of the power set of N, just as in the case when N is finite (Exercise 2.21 in the Cantor section). In particular, as mentioned earlier, the power of the continuum is equal to $2^{\aleph_0}$ (Exercises 2.6–2.7). We can now rephrase the claim Cantor made at the end of his 1878 paper about infinite linear sets. Clearly, the cardinal number of each infinite linear set will be greater or equal to $\aleph_0$ and less than or equal to the power of the continuum $\mathfrak{c}$. Furthermore,

$$\aleph_0 < 2^{\aleph_0} = \mathfrak{c}.$$

Cantor's assertion that there are exactly two cardinalities among the infinite subsets of the continuum now is equivalent to saying that $2^{\aleph_0}$ is exactly the next largest cardinal number after $\aleph_0$. This claim quickly became known as the *Continuum Hypothesis*, and its proof was of foremost importance to Cantor, who continued to work on it as long as he was mathematically active. The second and main source in this chapter is an excerpt from a long work that Cantor published in two installments in 1895 and 1897, entitled *Beiträge zur Begründung der Transfiniten Mengenlehre* (Contributions to the Founding of Transfinite Set Theory) [28, pp. 282–351]. In it he gives a grand summary and generalization of his earlier theory.

Another of Cantor's creations was to provide a new perspective on this problem, namely his theory of *ordinal numbers*. To create the cardinal number of a set, we are supposed to abstract from the nature of its elements as well as from the order in which they are given. If we now abstract only from the former, we obtain the *order type* of the set. Call two sets A and B, each equipped with a specific ordering, equivalent if there exists a one-to-one correspondence ϕ between their elements, which respects the order relation. Let us call such an equivalence an *order equivalence*. That is, if $a < a'$ in A, then $\phi(a) < \phi(a')$ in B. Naturally, a given set can usually be ordered in several different ways. For instance, the set $\mathbf{N}$ of natural numbers comes equipped with its natural ordering, but one could also order it by reversing the ordering, that is, $1 > 2 > 3 > \cdots$. Then one can show that there is no order-preserving one-to-one correspondence between these two ordered sets.

As another example, consider the set Q of rational numbers between 0 and 1. It, too, is ordered by its natural ordering. Cantor gave another, very interesting, ordering for Q [28, pp. 296 f.]. Let p/q and r/s be elements of Q, and suppose that they are written in reduced form, that is, the numerator and denominator have no common factors. Define

$$p/q \prec r/s$$

if either $p + q < r + s$ in the natural ordering of the integers, or $p + q = r + s$ and $p/q < r/s$ in the natural ordering of the rational numbers. Then one can show that with this new ordering, Q is order equivalent to $\mathbf{N}$ (Exercise 2.8). Furthermore, no such order equivalence can exist by using the natural ordering on Q. And, incidentally, this order equivalence shows that Q is countable. (Can this equivalence be extended to one between the set $\mathbf{Q}$ of all rational numbers and $\mathbf{N}$?)

Among ordered sets there is a special, distinguished kind, namely the so-called *well-ordered* sets. They have the property that any nonempty subset of a well-ordered set has a least element. The first infinite such set is, of course, $\mathbf{N}$. The set Q above with the ordering $\prec$ is well-ordered, even though it is not well-ordered with its natural ordering. (Why?) Thus, a given ordered set may not be well-ordered, but sometimes one can find a different ordering that is a well-ordering. One of the questions that will very soon enter our discussion asks whether this is *always* possible. Can every set be well-ordered? Well-ordered sets should be thought of as particularly simple and reminiscent of the ordered set $\mathbf{N}$ of natural numbers. They were singled out by Cantor to be the basis of his theory of ordinal numbers. As a result, this theory appears much more familiar and like the theory of natural numbers to us than that of the cardinal numbers.

Cantor defines the ordinal numbers simply to be all the order types of well-ordered sets. We add two ordinal numbers A and B by defining $A + B$ to be the disjoint union $A \uplus B$ with the following ordering. For $x, y \in A \uplus B$, let $x < y$ if either both x and y are in A, resp. B, and $x < y$ in the ordering on A, resp. B, or if $x \in A$ and $y \in B$. That is, every element of B is larger than every element of A. For example, if $A = \mathbf{N}$, and $B = \{a < b < c\}$, then $A + B$ is the order type

$$1 < 2 < 3 < \cdots < a < b < c.$$

Note that this order type is clearly different from the order type of $\mathbf{N}$, which is commonly denoted by ω (Exercise 2.9). Thus, we see that

$$\omega < \omega + 3.$$

If, on the other hand, we form $B + A$, then we obtain the order type of

$$a < b < c < 1 < 2 < \cdots,$$

which is easily seen to be the same order type as ω, that is,

$$3 + \omega = \omega.$$

Addition of ordinal numbers is therefore not commutative. Cantor also defined multiplication of ordinal numbers, which we shall not discuss here.

In one essential way, ordinal numbers are much easier to understand than cardinal numbers. Namely, it is not hard to prove that any two ordinal numbers can be compared [63, p. 254]. In other words, the trichotomy principle encountered earlier holds for ordinal numbers. In this way, they form a natural extension of the whole numbers:

$$1, 2, 3, \ldots, \omega, \omega + 1, \omega + 2, \ldots, \omega \cdot 2, \omega \cdot 2 + 1, \ldots.$$

It is time to return to the cardinal numbers, our main topic of interest. To every ordinal number α, we can now of course associate its cardinal number by taking any set of ordinal type α, which then has a cardinal number associated to it. Since any two sets of the same ordinal type also have the same cardinal type, it does not matter which set we take. This association is not unique (that is, not one-to-one), since different ordinal numbers may well have the same cardinal number, such as ω and $\omega + 1$, for instance. In this way we obtain a special class of cardinal numbers, namely those of well-ordered sets. And we can use certain ordinal numbers to label these cardinal numbers as follows. If we begin with the first infinite ordinal number ω, then its associated cardinal number is $\aleph_0$, which we have already encountered. Moving along in the series of ordinal numbers, at some point we encounter an ordinal that has a cardinal number larger than $\aleph_0$. (It is true, but not immediately obvious, that we will encounter a first such ordinal.) Call its cardinal number $\aleph_1$. Continuing in this way, we obtain a sequence of cardinal numbers

$$\aleph_0, \aleph_1, \aleph_2, \ldots, \aleph_\omega, \aleph_{\omega+1}, \ldots.$$

In particular, to every $\aleph_\alpha$, there is a unique next aleph, namely $\aleph_{\alpha+1}$. In fact, one can show that $\aleph_{\alpha+1}$ is the cardinal number of the well-ordered set of all ordinal numbers whose cardinal number is less than or equal to $\aleph_\alpha$. In particular, this set has larger cardinality than $\aleph_\alpha$. For details see [63, Ch. III, §11].

For the alephs, we have therefore verified the trichotomy principle. But what about the rest of the cardinal numbers? Cantor firmly believed that every cardinal number was in fact an aleph. Equivalently, he believed that every set could be well-ordered. In particular, he believed that the set of real numbers could be well-ordered. The Continuum Hypothesis can then be restated as the claim that

$$2^{\aleph_0} = \aleph_1.$$

In fact, one can now formulate a Generalized Continuum Hypothesis, to claim that if α is any ordinal number, then

$$2^{\aleph_\alpha} = \aleph_{\alpha+1}.$$

At the end of this Herculean construction effort in 1884, two results still eluded Cantor. First, a proof of the trichotomy principle was needed, or, equivalently, a proof that every cardinal number is an aleph, so that consequently every set can be well-ordered. This would provide the final validation of his cardinal numbers as a bona fide number system. Secondly, despite enormous effort, he was unable to prove the Continuum Hypothesis.

While a number of prominent mathematicians realized the magnitude of Cantor's creation, a large segment of the mathematical community viewed his transfinite numbers with skepticism or even outright hostility. This opposition adversely affected his professional and personal life, and might have contributed to a series of nervous breakdowns, which began in 1884 and were to continue for the rest of his life. He repeatedly decided to end his mathematical career, only to return to research, but with diminished power and creativity. Gradually, however, his work gained increased acceptance, and by the turn of the century Cantor's work was widely recognized as revolutionary and of extreme importance. During the Second International Congress of Mathematicians in Paris in 1900, David Hilbert (1862–1943), one of the great mathematicians of both the nineteenth and twentieth centuries, gave a celebrated lecture in which he outlined twenty-three problems, which in his opinion were the major unsolved problems for the new century [42]. As the very first one he listed the Continuum Hypothesis, together with the search for an explicit well-ordering of the real numbers. Anybody who solves one of Hilbert's problems today can expect immediate fame.

By the time the Third International Congress was to be held in Heidelberg, Germany, in 1904, there was growing concern, however, over several antinomies[2] that had been discovered and that threatened the very foundations on which set theory was built. Cantor had in fact observed certain paradoxical phenomena himself as early as 1895. In his attempts to show that every cardinal number was an aleph, he was led to consider the set of all sets. He had shown earlier that the power set of any set had to have larger cardinality than the set itself. This forced the conclusion that the set of all sets had to contain a set of larger cardinality, namely its power set. But this clearly contradicted the most basic results about cardinal numbers that Cantor had proved. The only possible conclusion was that the "set of all sets" could not be regarded as a completed, self-contained object. It was an "absolute infinite" that was beyond intellectual contemplation. This discovery did not worry Cantor particularly; he felt that it simply illuminated the limits of his theory, and indeed the limits of what was subject to rational investigation. Shortly thereafter, similar paradoxes were discovered by other mathematicians, the most startling of which was found by the British philosopher and logician Bertrand Russell (1872–1970). (See [87, pp. 124 f.] for the letter in which Russell communicates the paradox to the German mathematician Gottlob Frege (1848–1925).) His paradox pointed to an even more basic problem inherent in the very fundamentals of set theory. His construction was essentially a modern version of the classical "barber paradox," according to which there is a town with a (male) barber who shaves precisely all those men in the town who do not shave themselves. Then the question is posed whether the barber shaves himself. Either a "yes" or a "no" answer to the question quickly leads to a contradiction. (Try it!) Russell's version of this paradox was the set whose elements were exactly all those sets that do not contain themselves as an element. Here, too, either answer to the question whether this set contains itself as an element leads to a contradiction (Exercise 2.10). Set theory clearly had an inherent problem, and everyone was getting increasingly worried.

But things were about to get worse for Cantor. During the Heidelberg Congress, he attended a talk of the Hungarian mathematician Julius König (1849–1914),

---

[2]Statements that contradict themselves.

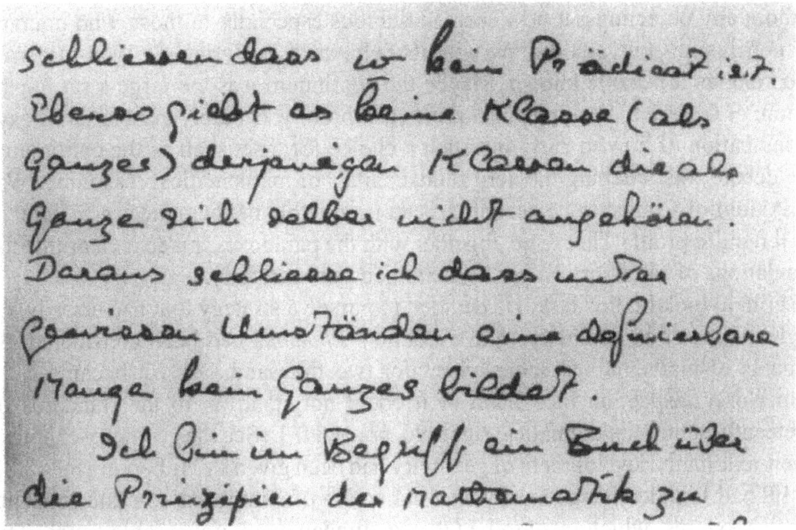

schliessen dass w kein Prädicat ist.
Ebenso giebt es keine Klasse (als
Ganzes) derjenigen Klassen die als
Ganze sich selber nicht angehören.
Daraus schliesse ich dass unter
gewissen Umständen eine definierbare
Menge kein Ganzes bildet.
Ich bin im Begriff ein Buch über
die Prinzipien der Mathematik zu

PHOTO 2.1. Russell's letter to Frege.

who claimed to have a proof that the cardinal number of the continuum was not an aleph. Consequently, the real numbers could not be well-ordered. A central pillar of Cantor's belief system appeared to be shattered. Cantor was deeply humiliated and furious, his work and reputation challenged before the assembled mathematical community. Less than twenty-four hours later, the young German Ernst Zermelo (1871–1953) had found a flaw in König's proof. While an immediate catastrophe had been averted, the seeds of doubt were irreversibly sown in Cantor and everybody else. He felt that there was a real possibility now that somebody could find a way to seriously undermine the plausibility of the Continuum Hypothesis. After the Congress, a group of mathematicians, including Hilbert, gathered to discuss the implications of König's paper. Even though it was clear by the end of the Congress that König's proof did not hold water, nobody could be sure that it was not possible to patch it up.

Then, a month later, Zermelo sent an excited letter to Hilbert outlining a proof that *every* set could be well-ordered, and therefore König's proof was impossible to fix. Zermelo's result implied also, of course, a proof that every cardinal number was an aleph, hence the trichotomy principle for general cardinal numbers was proven, since alephs had already been shown to be comparable. In particular, the cardinal number of the continuum was an aleph. Now it was just a matter of deciding which one. All was well, or so it seemed. After the speedy publication of Zermelo's proof and careful scrutiny by his peers all over Europe, criticism focused on one particular step. For a given set S, Zermelo assumed that it was possible to simultaneously choose an element from every nonempty subset of S. The assumption that this is always possible is known as the *Axiom of Choice*. And while it was pointed out that this axiom was being used tacitly all over mathematics without any objections, it now seemed dubious especially to those who doubted the conclusions one could draw with its help, such as Zermelo's Well-Ordering Theorem, as it became known. Maybe the continuum was too large a set for the Axiom of Choice to be applicable. Zermelo's proof was purely existential; it gave no indication of how to carry out such a choice for a set such as the continuum. The debate was touching the very fundamentals of mathematical reasoning. Was the Axiom of Choice to be accepted as an irreducible mathematical principle, or did it require proof? This issue, together with the paradoxes, made it clear that the foundations of set theory had to be examined very carefully.[3]

Zermelo took on this task. He decided to pursue a strategy that had already led to great success in geometry. While Euclid's *Elements* had been considered the pinnacle of mathematical achievement for two thousand years, it became clear upon closer inspection that much of it could not stand up to the standards of nineteenth century mathematical rigor [96, pp. 188 ff.]. (See the geometry chapter.) A new axiomatic development of geometry had been given by Hilbert in 1899 [88]. Euclid had founded his geometry on a collection of definitions of basic concepts and five axioms, which were to be taken as fundamental, unprovable truths. The whole structure of Euclidean geometry was to unfold from this basis by logical deduction. One problem with Euclid's definitions is, of course, that in order to define a concept, other concepts are needed, which would then need to be defined in turn, leading to an infinite chain of definitions. Hilbert solved this problem by simply leaving notions like "point" and "line" undefined, specifying only their relationships.

Similarly, Zermelo did not attempt to define the basic notion of "set" and "element of," which seemed to lie at the heart of most of the paradoxes. By 1907 Zermelo had worked out a list of seven axioms, which were to allow the rigorous development of Cantor's set theory, including Zermelo's Well-Ordering Theorem, while at the same time excluding all known paradoxes. Basically, the axioms specified the existence of certain sets and a number of well-defined procedures to build new sets from old ones. These procedures were to be sufficiently powerful to construct all sets needed in the Cantorian theory, but not powerful enough to construct sets large enough to lead to trouble, like Russell's paradox. Zermelo published his article *Untersuchungen über die Grundlagen der Mengenlehre, I* (Investigations on the Foundations of Set Theory, I) in the journal *Mathematische Annalen* in 1908. The last original source is an excerpt from this paper including his list of axioms.

---

[3] While the Axiom of Choice might appear quite reasonable and harmless at first sight, one can actually prove rather amazing facts with it, beyond the well-ordering of sets. For instance, it was proven by the two Polish mathematicians Stefan Banach and Alfred Tarski that one can take a ball in three-dimensional space, cut it up into finitely many congruent pieces, and then reassemble the pieces to obtain a ball of *arbitrarily large* volume. One can furthermore show that this result cannot be proven without the use of the Axiom of Choice. (See [176].)

While his paper represented a big step in the right direction, Zermelo had not entirely succeeded in addressing all important issues. He could not provide a proof that the axioms were consistent, that is, did not allow the proof of contradictory theorems, and he could not show that they were independent, that is, that none of them could be proven from the others. However, with some subsequent modifications by the German Adolf Fraenkel (1891–1965) and others, the system known as Zermelo–Fraenkel set theory is now the most commonly accepted model for set theory. We shall subsequently refer to this system as **ZFC** to emphasize the special role of the Axiom of Choice, while **ZF** shall denote the system without the Axiom of Choice.

This state of affairs persisted for more than twenty years, despite great efforts by some of the best mathematical minds to resolve the issues of consistency and independence. Then, like a bombshell, news of a most disturbing result hit the mathematics community. The Austrian mathematician Kurt Gödel (1906–1978), destined to become the greatest logician since Aristotle, announced a result at a conference in 1930, now widely known as his *Incompleteness Theorem*, that implied that it was impossible to prove the consistency of the axioms of set theory within that theory. That could only be done in a larger, higher-order theory, which was then in turn subject to questions about its own consistency. With this result, Gödel, who had just received his Ph.D. the year before, doomed all efforts to provide absolute proof of the consistency of set theory, and any other axiomatic theory that contained number theory within it (see the analysis chapter). This negative result was followed in 1935 by a proof that the Axiom of Choice is relatively consistent with the other axioms of **ZFC**. That is, if one could prove a contradiction in **ZFC**, then it could already be proved in **ZF**. And then, in 1939, he supplied a proof that the Continuum Hypothesis, even its generalized form, was consistent with **ZFC**; that is, if one could prove a contradiction using the axioms of **ZFC** together with the Continuum Hypothesis, then this contradiction could already be proved using the axioms alone. Gödel had been in possession of these results since 1937, but had delayed publication in the hope of finding a proof that the Continuum Hypothesis was in fact *independent* of **ZFC**, that is, that the axioms of **ZFC** were not strong enough either to prove or disprove it [42]. But his persistent efforts to substantiate his belief were unsuccessful, even though he continued to work on it for another decade. (A rather unusual view of Gödel's work is contained in [89].)

Such a proof had to wait until 1963, when Stanford mathematician Paul Cohen proved that if **ZFC** is consistent, then it remains consistent if one adds the negation of the Continuum Hypothesis. He also showed that the Axiom of Choice is independent of the **ZF** axioms. While this celebrated result provided a temporary ending to almost a century of efforts to prove the Continuum Hypothesis, it is of course a most unsatisfactory one. Most mathematicians are Platonists, in the sense that they believe the objects of mathematics to be real, rather than just the pieces in a formal game with symbols. And if the objects of set theory are considered real, then the Continuum Hypothesis is either true or false. Consequently, the appropriate way to continue is to search for other axioms considered *true*, which,

PHOTO 2.2. Gödel.

when added to **ZFC**, allow either a proof or a disproof of the Hypothesis. This was Gödel's point of view, and he believed that the Continuum Hypothesis would turn out to be false. Many set theorists today agree. (See [124, 125] for a history of the Continuum Hypothesis and the Axiom of Choice. A good account of the paradoxes of early set theory can be found in [62], and a very good description of Cantor's set theory is in [63].)

Whatever the outcome, the work of Cantor in general and the Continuum Hypothesis in particular has forever changed the face of mathematics. Set theory is at the basis of all mathematics; the fields of mathematical logic, so essential in the design of computers, and topology, widely used in analysis and geometry, are direct outgrowths of the tools developed to solve the problems posed by Cantor's work.

Exercise 2.1: Look up the other three paradoxes of Zeno on the impossibility of motion and explain his reasoning. Can you refute any of them?

Exercise 2.2: Find a one-to-one correspondence between the set of all integers and that of all positive integers.

Exercise 2.3: Show that the points on any two line segments of arbitrary finite length are in one-to-one correspondence.

Exercise 2.4: Show that the notion of one-to-one correspondence of sets satisfies the properties of an equivalence relation, namely that it is reflexive, transitive, and symmetric.

Exercise 2.5: Show that there is a one-to-one correspondence between the subsets of the set **N** of natural numbers and the set of functions from **N** into $\{0, 1\}$. Use the binary expansion of real numbers to show that there is a one-to-one correspondence between the set **R** of real numbers and the power set of **N**.

Exercise 2.6: Show that if N is a cardinal number, then $N^1 = N$, and $1^N = 1$. If c is the cardinal number of the continuum, show that $c^{\aleph_0} = c$.

Exercise 2.7: Extend the product of finitely many cardinal numbers to infinitely many factors. Show that $1 \cdot 2 \cdot 3 \cdot 4 \cdots = c$.

Exercise 2.8: Using the ordering $\prec$ that Cantor gives for the set Q of rational numbers between 0 and 1, give an explicit one-to-one correspondence between Q and the natural numbers **N**.

Exercise 2.9: If α is the order type of an ordered set S, let α^* denote the order type of the set S^* that has the same elements as S, but with the reverse ordering. For instance, if $S = \{a < b\}$, then $S^* = \{a > b\}$. Give an example of an ordered set whose order type is $\omega^* + \omega$. (Addition of order types is defined in the same way as that of ordinal numbers.)

Exercise 2.10: Write out the details of Russell's paradox.

2.2   Bolzano's Paradoxes of the Infinite

"My special pleasure in mathematics rested particularly on its purely speculative parts; in other words, I prized only that part of mathematics which was at the same time philosophy" [178, p. 64] (see also [148, p. 46]). This philosophical bent to Bernard Bolzano's mathematical interests quite naturally attracted him to the study of foundational questions and the philosophy of the infinite.

Bolzano entered the University of Prague in 1796, where he studied philosophy, mathematics, and physics. He was especially inspired to study mathematics through the book *Foundations of Mathematics* by Abraham Kästner, a professor at the University of Göttingen, Germany, saying of it: "He proved what is generally passed over because everyone already knows it, i.e., he sought to make the reader clearly aware of the basis on which his judgements rest. That was what I liked most of all" [178, p. 64] [148, p. 46].

As mentioned in the introduction, Bolzano chose a career as a theologian over mathematics. At that time Bohemia, of which Prague was the major city, was part of the Habsburg Empire ruled by Austria. Emperor Franz I decided in 1805 that each university should have a new chair in the philosophy of religion. His motivation was political: he hoped to curb the spread of freethinking taking place in Bohemia following the French Revolution. Part of Bolzano's duties was to give twice-weekly sermons. His lectures became widely popular, sometimes drawing an audience of

PHOTO 2.3. Bolzano.

more than a thousand listeners. The content of his speeches, however, ran quite contrary to the intentions of Emperor Franz. Bolzano espoused socialist and pacifist views and criticized religious doctrine. Not surprisingly, he was dismissed by imperial order in 1819, forbidden to hold any teaching position, and barred from publishing. With the help of friends, Bolzano was able to spend the rest of his life working mostly on two projects, his *Theory of Science* [15], dealing with logic, and the *Theory of Magnitude*, on mathematics. The latter was intended to be a systematic development of all of mathematics, but unfortunately remained unfinished. More about Bolzano's life can be found in [42, 148].

In today's mathematics, Bolzano's name appears most frequently in analysis, in connection with the so-called Bolzano–Weierstrass Theorem. As mentioned earlier, a major task for nineteenth-century mathematicians was to build a solid foundation for analysis, the theory of continuous functions and infinite series, and Bolzano made some significant contributions to this program. His approach was quite similar to that successfully carried out later by mathematicians like Weierstrass, Dedekind, and Cantor. It can be described as follows:

> In his work and, indeed, in his broader programme, Bolzano did not seek merely to "arithmetize analysis," that is, he did not wish only to purge the concepts of limit, convergence, and derivative of geometrical components and replace them by purely arithmetic concepts. He was aware of a deeper problem: the need to refine and enrich the concept of number itself.

It is within this broader framework that the importance and effectiveness of Bolzano's definitions of convergence and continuity can best be understood [148, p. 49].

In the paper *Purely Analytic Proof of the Theorem That Between Any Two Values Which Give Results of Opposite Sign There Lies at Least One Real Root of the Equation*, published by Bolzano in 1817 (see [147] for an English translation), he gives the first rigorous proof of what is now known as the important Intermediate Value Theorem in calculus:

> If two functions of x, $f(x)$ and $\phi(x)$, vary *according to the law of continuity* either for *all* values of x or only for those which lie between α and β, and if $f(\alpha) < \phi(\alpha)$ and $f(\beta) > \phi(\beta)$, then there is always a certain value of x between α and β for which $f(x) = \phi(x)$ [147, p. 177].

In order to prove this theorem, Bolzano needed to know a crucial property of the real numbers. He states and proves the following early form of the Bolzano–Weierstrass Theorem:

> If a property M does not belong to *all* values of a variable x, but does belong to *all* values which are *less* than a certain u, then there is always a quantity U which is the greatest of those of which it can be asserted that all smaller x have property M [147, p. 174].

Weierstrass's name became attached to the result when he later generalized it from the continuum to higher-dimensional spaces. Augustin-Louis Cauchy (1789–1857) in France proved results about infinite series that paralleled those Bolzano had obtained, apparently unaware of Bolzano's results.

Much of Bolzano's work after his dismissal centered on a major effort, the *Theory of Magnitude*. Although he never completed this work, his friend Prihonsky edited and published a part of this project after Bolzano's death, the book *Paradoxien des Unendlichen* (Paradoxes of the Infinite), which has become Bolzano's best-known work. Here we present excerpts in which Bolzano argues quite convincingly that the most promising approach toward an analysis of the infinite is mathematical, and observes properties of infinite sets that point in a direction later followed by Cantor. The translation is made from the German edition [17, pp. 28–31].

Bernard Bolzano, from
Paradoxes of the Infinite

§19

The examples of infinity considered so far already make it clear that not all infinite sets can be considered *equal with respect to their multiplicity*. Instead, some are *larger* (or *smaller*) than others, that is, one can encompass the other as a mere part (or, conversely, one can be a mere part in another). This too is a claim which sounds *paradoxical* to many. And of course those who define the infinite as something that is not capable of being enlarged must find it not only paradoxical but *contradictory* that one infinite should be larger than another. But we have already found above that this view is based on a concept of the infinite which does not conform with the usage of the word. Our explanation, which does not only conform to common usage of the word, but is also in accordance with the purpose of science, cannot allow anyone to find something contradictory, not even something remarkable, in the thought that one infinite set should be larger than another.

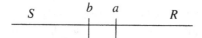

Who, for instance, would not agree that the length of the line which is unlimited in the direction aR is infinite? And that the line from b in the same direction bR must be considered greater than aR by the piece ba? And that the line which extends without limit in both directions aR and aS is larger still, by a quantity, which is infinite itself? Etc.

§20

Let us move on now to consider a most remarkable peculiarity which can appear in the relationship between two sets if *both are infinite*. In fact, it always appears, but has so far been overlooked, to the disadvantage of the discovery of a number of important truths in metaphysics, as well as in physics and mathematics. As I state them, they will be found so paradoxical even now, that it might be necessary to devote some time to their contemplation. Namely, I claim: two sets which are both infinite can be related in such a way that *on the one hand*, each element of one set can be combined with an element of the other set in such a way that no element of either set is left without being paired up, and that no element appears in two or more pairs. *On the other hand*, it is possible that one of these sets contains the other one as a mere *part*, so that the magnitudes which they represent, when we consider all elements as identical, that is, as units, can be in *manifold different relationships*.

I will prove this claim via two examples, in which the assertion undeniably appears.

1. If we take two arbitrary (abstract) numbers, 5 and 12, for instance, then it is apparent that the set of numbers between 0 and 5 (or those which are smaller than 5) is infinite, just as the set of numbers less than 12. But it is just as apparent that the latter set is larger than the former, since the former is undeniably only a part of the latter. If we replace the numbers 5 and 12 by any two others, then we are forced to conclude that the two sets don't always retain the same relationship, but can be related in a variety of ways. But the following is equally true: If x is any number between 0 and 5, and if we determine the relationship between x and y by the equation

$$5y = 12x,$$

then y is a number between 0 and 12. Conversely, whenever y is between 0 and 12, x will lie between 0 and 5. The equation furthermore implies that to every value of x is related only one value of y, and vice versa. The two facts make it clear that for every number x between 0 and 5, there is a number y between 0 and 12 which can be paired up with it in such a way that not a single element in the two sets remains unrelated to a pair, and that not a single one appears in two or more pairs.

...

§21

Therefore, we are not allowed to conclude that two sets A and B are *equal to one another* in regard to the multiplicity of their parts (that is, if we disregard all their distinct features) *if they are infinite*, just because they are related to each other in such a way that for each part a of A, one can find, according to a certain rule, a part b of B such that the collection of pairs (a, b) formed in this way contains each element of A or B once and only once. Despite this relationship, which in itself gives an equivalence between them, there can be a relationship of inequality of their multiplicities, so that one can be a whole, of which the other is a part. To conclude the equality of these sets, an additional reason is necessary, such as, that both sets are determined in the exact same way, for instance, or are created in the same way.

Bolzano presents his ideas very clearly, so little annotation is needed to make the text understandable. However, a few minor points deserve comment. In Section 20 Bolzano gives examples of sets that are in one-to-one correspondence with a proper subset, just as Galileo did. Bolzano, however, concluded in Section 21 that this phenomenon was indeed intrinsic to infinite sets (it in fact leads to one possible definition of infinite set) and a more elaborate quantitative comparison of infinite sets is not precluded. In fact, a good part of the book elaborates on the apparently paradoxical relationship of infinite sets with proper subsets, and shows that it must be accorded a central role in the whole theory. Just such a comparison of sizes of infinite sets was undertaken by Georg Cantor, with considerable mathematical rigor.

Exercise 2.11: In Example 1 of §20, Bolzano shows that the set of real numbers y between 0 and 12 can be put in one-to-one correspondence with a proper subset of itself, namely the numbers x between 0 and 5, by means of the equation

$$5y = 12x.$$

Verify that $25y = 12x^2$ and $5y = 12(5 - x)$ will also give such one-to-one correspondences. Can you give additional examples?

Exercise 2.12: Show that the points of a 1 inch by 1 inch square can be put into one-to-one correspondence with the points of a 2 inch by 2 inch square and with the points of a 2 inch by 1 inch rectangle.

Exercise 2.13: Give additional examples of sets that may be put in one-to-one correspondence with proper subsets of themselves.

2.3   Cantor's Infinite Numbers

On the occasion of Georg Cantor's confirmation, his father wrote a letter to him that Cantor would always remember. It foreshadowed much that was to happen to Cantor in later years.

> No one knows beforehand into what unbelievably difficult conditions and occupational circumstances he will fall by chance, against what unforeseen and unforeseeable calamities and difficulties he will have to fight in the various situations of life.

> How often the most promising individuals are defeated after a tenuous, weak resistance in their first serious struggle following their entry into practical affairs. Their courage broken, they atrophy completely thereafter, and even in the best case they will still be nothing more than a so-called ruined genius!...

> But they lacked that steady heart, upon which everything depends! Now, my dear son! Believe me, your sincerest, truest and most experienced friend—this sure heart, which must live in us, is: a truly religious spirit....

> But in order to prevent as well all those other hardships and difficulties which inevitably rise against us through jealousy and slander of open or secret enemies in our eager aspiration for success in the activity of our own specialty or business; in order to combat these with success one needs above all to acquire and to appropriate the greatest amount possible of the most basic, diverse technical knowledge and skills. Nowadays these are an absolute necessity if the industrious and ambitious man does not want to see himself pushed aside by his enemies and forced to stand in the second or third rank [36, pp. 274 ff.].

By the time Cantor encountered his own great "unforeseen calamities and difficulties" he had acquired that religious spirit and steady heart that his father believed to be so indispensable.

Having been irresistably drawn to mathematics, Cantor excelled in it as a student, without neglecting diversity in learning, as his father had admonished him. In 1869, at age 22, three years after he completed his dissertation at the University of Berlin, he joined the University of Halle as a lecturer. Through his colleague Eduard Heine (1821–1881), Cantor became interested in analysis, and soon drew the attention of important researchers, most notably Leopold Kronecker (1823–1891), a professor at the University of Berlin and a very influential figure in German mathematics, both mathematically as well as politically. Kronecker held very strong views about the admissibility of certain mathematical techniques and approaches. To him is attributed the statement "God Himself made the whole numbers—everything else is the work of men." As a consequence, he disliked the style of analysis done by Weierstrass, involving a concept of the real numbers that was not based on whole numbers, and which was not constructive to a high degree. As soon as Cantor began to pursue his set-theoretic research, he lost the support of Kronecker, who quickly became one of his strongest critics. Cantor's desire to obtain a position in the

PHOTO 2.4. Cantor with his wife, Vally.

mathematically much more stimulating environment of Berlin never materialized, and Kronecker's opposition had much to do with it.

The series of six articles on infinite linear sets that Cantor published between 1878 and 1884 contained in essence the whole of his set-theoretic discoveries. His creative period was close to an end. Shortly thereafter, in the spring of 1884, he suffered his first mental breakdown. But by the fall he was back working on the Continuum Hypothesis. An important supporter of Cantor in those years was the Swedish mathematician Gösta Mittag-Leffler (1846–1927). A student of Weierstrass, Mittag-Leffler became professor at the University of Stockholm in 1881. Shortly afterwards he founded the new mathematics journal *Acta Mathematica*, which quickly became very influential, with articles by the most eminent researchers. His relationship with Cantor was close, and Mittag-Leffler offered to publish Cantor's papers when other journals were rather reluctant to do so. In 1885, Cantor was about to publish two short articles in the *Acta Mathematica* containing some new ideas on ordinal numbers. While the first article was being typeset, Cantor received the following letter from Mittag-Leffler, requesting that Cantor withdraw the article:

> I am convinced that the publication of your new work, before you have been able to explain new positive results, will greatly damage your reputation among mathematicians. I know very well that basically this is all the same

to you. But if your theory is once discredited in this way, it will be a long time before it will again command the attention of the mathematical world. It may well be that you and your theory will never be given the justice you deserve in your lifetime. Then the theory will be rediscovered in a hundred years or so by someone else, and then it will subsequently be found out that you already had it all. Then, at least, you will be given justice. But in this way [by publishing the article], you will exercise no significant influence, which you naturally desire as does everyone who carries out scientific research [36, p. 138].

The effect on Cantor was devastating. Disillusioned and without friends in the mathematical community, he decided by the end of 1885 to abandon mathematics. Instead, he turned to philosophy, theology, and history. To his surprise, he found more support for his work among theologians than he had ever received from mathematicians. Soon, however, he was back at work on an extension and generalization of his earlier work. As his last major set-theoretic work, he published a grand summary of his set theory between 1895 and 1897, his *Beiträge zur Begründung der Transfiniten Mengenlehre* (Contributions to the Founding of Transfinite Set Theory) [28, 282–351]. The following excerpt from this work concerns the arithmetic of cardinal numbers. The English translation is taken from [26], pp. 85–97, 103–108.

In the wake of this publication, his work was finally recognized for its full significance, and while controversy continued to surround set theory, Cantor was not alone anymore to defend it. Set theory was here to stay. His mathematical career had reached its zenith, and he was able to see that he had indeed created something of lasting significance.

Georg Cantor, from

Contributions to the Founding of the Theory of Transfinite Numbers

§1
The Conception of Power or Cardinal Number

By an "aggregate"[4] we are to understand any collection into a whole M of definite and separate objects m of our intuition or our thought. These objects are called the "elements" of M.

In signs we express this thus:

(1) $$M = \{m\}.$$

---

[4]The modern terminology is "set."

We denote the uniting of many aggregates $M, N, P, \ldots$, which have no common elements, into a single aggregate by

(2)
$$(M, N, P, \ldots).$$

The elements of this aggregate are, therefore, the elements of M, of N, or $P, \ldots$, taken together.

We will call by the name "part" or "partial aggregate" of an aggregate M any other aggregate M_1 whose elements are also elements of M.

If M_2 is a part of M_1 and M_1 is a part of M, then M_2 is part of M.

Every aggregate M has a definite "power," which we will also call its "cardinal number."

We will call by the name "power" or "cardinal number" of M the general concept which, by means of our active faculty of thought, arises from the aggregate M when we make abstraction of the nature of its various elements m and of the order in which they are given.

We denote the result of this double act of abstraction, the cardinal number or power of M, by

(3)
$$\overline{\overline{M}}.$$

Since every single element m, if we abstract from its nature, becomes a "unit," the cardinal number $\overline{\overline{M}}$ is a definite aggregate composed of units, and this number has existence in our mind as an intellectual image or projection of the given aggregate M.

We say that two aggregates M and N are "equivalent," in signs

(4)
$$M \sim N \text{ or } N \sim M,$$

if it is possible to put them, by some law, in such a relation to one another that to every element of each one of them corresponds one and only one element of the other. To every part M_1 of M there corresponds, then, a definite equivalent part N_1 of N, and inversely.

If we have such a law of co-ordination of two equivalent aggregates, then, apart from the case when each of them consists only of one element, we can modify this law in many ways. We can, for instance, always take care that to a special element m_0 of M a special element n_0 of N corresponds. For if, according to the original law, the elements m_0 and n_0 do not correspond to one another, but to the element m_0 of M the element n_1 of N corresponds, and to the element n_0 of N the element m_1 of M corresponds, we take the modified law according to which m_0 corresponds to n_0 and m_1 to n_1 and for the other elements the original law remains unaltered. By this means the end is attained.

Every aggregate is equivalent to itself:

(5)
$$M \sim M.$$

If two aggregates are equivalent to a third, they are equivalent to one another; that is to say:

(6) $\qquad$ from $M \sim P$ and $N \sim P$ follows $M \sim N$.

Of fundamental importance is the theorem that two aggregates M and N have the same cardinal number if, and only if, they are equivalent: thus,

(7) $\qquad$ from $M \sim N$ we get $\overline{\overline{M}} = \overline{\overline{N}}$,

and

(8) $\qquad$ from $\overline{\overline{M}} = \overline{\overline{N}}$, we get $M \sim N$.

Thus the equivalence of aggregates forms the necessary and sufficient condition for the equality of their cardinal numbers.

In fact, according to the above definition of power, the cardinal number $\overline{\overline{M}}$ remains unaltered if in the place of each of one or many or even all elements m of M other things are substituted. If, now, $M \sim N$, there is a law of co-ordination by means of which M and N are uniquely and reciprocally referred to one another; and by it to the element m of M corresponds the element n of N. Then we can imagine, in the place of every element m of M, the corresponding element n of N substituted, and, in this way, M transforms into N without alteration of cardinal number. Consequently

$$\overline{\overline{M}} = \overline{\overline{N}}.$$

The converse of the theorem results from the remark that between the elements of M and the different units of its cardinal number $\overline{\overline{M}}$ a reciprocally univocal (or bi-univocal) relation of correspondence subsists.[5] For, as we saw, $\overline{\overline{M}}$ grows, so to speak, out of M in such a way that from every element m of M a special unit of $\overline{\overline{M}}$ arises. Thus we can say that

(9) $\qquad\qquad M \sim \overline{\overline{M}}.$

In the same way $N \sim \overline{\overline{N}}$. If then $\overline{\overline{M}} = \overline{\overline{N}}$, we have, by (6), $M \sim N$.

We will mention the following theorem, which results immediately from the conception of equivalence. If $M, N, P, \ldots$ are aggregates which have no common elements, $M', N', P', \ldots$ are also aggregates with the same property, and if

$$M \sim M', \ N \sim N', \ P \sim P', \ \ldots,$$

then we always have

$$(M, N, P, \ldots) \sim (M', N', P', \ldots).$$

Early on in this section Cantor defines the concept of *cardinal number* of an aggregate (or set, in modern parlance) which, in his words, is "the general concept which, by means of our active faculty of thought, arises from the aggregate

---

[5] One-to-one correspondence.

M when we make abstraction of the nature of its various elements m and of the order in which they are given." Cantor then observes that the result $\overline{\overline{M}}$ of these abstractions is again a set. Thus Cantor employs a process of abstraction to define a concept based on a common feature shared by a collection of objects. For instance, the concept "blue" is abstracted from that quality which all blue objects have in common, and "three" arises from the quality that all sets with three elements have in common. In particular, Cantor notes that not only does the abstraction of the quality of having three elements have nothing to do with the particular nature of the elements, it is also independent of the order in which they are given. However, the reader presumably is struck by the seeming circularity of this definition.

The confusion stems from the subtle fact that the first occurrence of "three" in the above sentence is a noun to be defined through the adjective "three." A slightly different formulation of the same idea is to say that the noun "three" denotes that property which all sets have in common that can be put in one-to-one correspondence with the set $\{a, b, c\}$. (In fact, in §5, Cantor gives roughly this definition for the finite cardinal number 3.) This second definition seems at first to avoid the problem of circularity, though it may seem a bit arbitrary. What is special about the set $\{a, b, c\}$? Why not use $\{X, Y, Z\}$? For that matter, why not use the set $\{x, y\}$? The obvious answers are that there is nothing special about $\{a, b, c\}$; $\{X, Y, Z\}$ would do just as well, since it also has three elements; but $\{x, y\}$ does not work, since it has only two elements. We have not escaped circularity after all. Both attempts to define "three" suffer from the same essential problem: they both rely on the idea that we can recognize whether or not a given set has three elements *before* we have a definition of three. The puzzle as to how we can recognize that some particular concrete object has an abstract property (i.e., that $\{a, b, c\}$ has three elements or that the Cookie Monster is blue) without having a definition of the property (three or blue) dates back at least as far as Plato and has an extensive literature.

A further logical problem with this way of defining cardinal numbers (or anything, for that matter) is that one might define a property and give it a name without being assured that it really exists, or that a unique object of this kind exists. (Bertrand Russell argues in [149, pp. 114–15] that definition by abstraction is never valid.) Later mathematicians and logicians objected to Cantor's approach and searched for a definition that would incorporate the existence of the thing so defined by giving the definition in terms of things already known to exist. Russell and, independently, Gottlob Frege around the turn of the century proposed to define the cardinal number of a set as the collection of all sets that could be put in one-to-one correspondence with it, thereby specifying a unique object. However, this definition also leads to serious logical problems, since any such collection is subject to Russell's paradox mentioned in the introduction, that is, it is too big to be a set and thereby lies outside the realm of a theory of sets free of paradoxes. This difficulty in turn led to a further refinement of the idea of cardinal number, through the use of ordinal numbers, proposed by John von Neumann in 1928 [125, p. 265], and has become the modern standard.

§2
"Greater" and "Less" with Powers

If for two aggregates M and N with the cardinal numbers $\mathfrak{a} = \overline{\overline{M}}$ and $\mathfrak{b} = \overline{\overline{N}}$, both the conditions:

(a) There is no part of M which is equivalent to N,
(b) There is a part N_1 of N, such that $N_1 \sim M$,

are fulfilled, it is obvious that these conditions still hold if in them M and N are replaced by two equivalent aggregates M' and N'. Thus they express a definite relation of the cardinal numbers $\mathfrak{a}$ and $\mathfrak{b}$ to one another.

Further, the equivalence of M and N, and thus the equality of $\mathfrak{a}$ and $\mathfrak{b}$, is excluded; for if we had $M \sim N$, we would have, because $N_1 \sim M$, the equivalence $N_1 \sim N$, and then, because $M \sim N$, there would exist a part M_1 of M such that $M_1 \sim M$, and therefore we should have $M_1 \sim N$; and this contradicts the condition (a).

Thirdly, the relation of $\mathfrak{a}$ to $\mathfrak{b}$ is such that it makes impossible the same relation of $\mathfrak{b}$ to $\mathfrak{a}$; for if in (a) and (b) the parts played by M and N are interchanged, two conditions arise which are contradictory to the former ones.

We express the relation of $\mathfrak{a}$ to $\mathfrak{b}$ characterized by (a) and (b) by saying: $\mathfrak{a}$ is "less" than $\mathfrak{b}$ or $\mathfrak{b}$ is "greater" than $\mathfrak{a}$; in signs

(1) $$\mathfrak{a} < \mathfrak{b} \text{ or } \mathfrak{b} > \mathfrak{a}.$$

We can easily prove that,

(2) $$\text{if } \mathfrak{a} < \mathfrak{b} \text{ and } \mathfrak{b} < \mathfrak{c}, \text{ then we always have } \mathfrak{a} < \mathfrak{c}.$$

Similarly, from the definition, it follows at once that, if P_1 is part of an aggregate P, from $\mathfrak{a} < \overline{\overline{P_1}}$ follows $\mathfrak{a} < \overline{\overline{P}}$ and from $\overline{\overline{P}} < \mathfrak{b}$ follows $\overline{\overline{P_1}} < \mathfrak{b}$.

We have seen that, of the three relations

$$\mathfrak{a} = \mathfrak{b}, \ \mathfrak{a} < \mathfrak{b}, \ \mathfrak{b} < \mathfrak{a},$$

each one excludes the two others. On the other hand, the theorem that, with any two cardinal numbers $\mathfrak{a}$ and $\mathfrak{b}$, one of those three relations must necessarily be realized, is by no means self-evident and can hardly be proved at this stage.

Not until later, when we shall have gained a survey over the ascending sequence of the transfinite cardinal numbers and an insight into their connexion, will result the truth of the theorem:

A. If $\mathfrak{a}$ and $\mathfrak{b}$ are any two cardinal numbers, then either $\mathfrak{a} = \mathfrak{b}$ or $\mathfrak{a} < \mathfrak{b}$ or $\mathfrak{a} > \mathfrak{b}$.

Here, Cantor states the trichotomy principle, comments that this assertion is by no means self-evident, and promises to provide a proof later. No use of the assertion is made in the text at hand. Later Cantor claimed in a letter to Dedekind [28, p. 447] to have a proof that relied on the Well-Ordering Theorem.

§3

The Addition and Multiplication of Powers

The union of two aggregates M and N which have no common elements was denoted in §1, (2), by (M, N). We call it the "union-aggregate of M and N."

If M' and N' are two other aggregates without common elements, and if $M \sim M'$ and $N \sim N'$, we saw that we have

$$(M, N) \sim (M', N').$$

Hence the cardinal number of (M, N) only depends upon the cardinal numbers $\overline{\overline{M}} = a$ and $\overline{\overline{N}} = b$.

This leads to the definition of the sum of a and b. We put

(1)
$$a + b = \overline{\overline{(M, N)}}.$$

Since in the conception of power, we abstract from the order of the elements, we conclude at once that

(2)
$$a + b = b + a;$$

and, for any three cardinal numbers a, b, c, we have

(3)
$$a + (b + c) = (a + b) + c.$$

We now come to multiplication. Any element m of an aggregate M can be thought to be bound up with any element n of another aggregate N so as to form a new element (m, n); we denote by $(M.N)$ the aggregate of all these bindings (m, n), and call it the "aggregate of bindings of M and N."[6] Thus

(4)
$$(M.N) = \{(m, n)\}.$$

We see that the power of $(M.N)$ only depends on the powers $\overline{\overline{M}} = a$ and $\overline{\overline{N}} = b$; for, if we replace the aggregates M and N by the aggregates

$$M' = \{m'\} \text{ and } N' = \{n'\}$$

respectively equivalent to them, and consider m, m' and n, n' as corresponding elements, then the aggregate

$$(M'.N') = \{(m', n')\}$$

is brought into a reciprocal and univocal correspondence with $(M.N)$ by regarding (m, n) and (m', n') as corresponding elements. Thus

(5)
$$(M'.N') \sim (M.N).$$

We now define the product $a.b$ by the equation

(6)
$$a.b = \overline{\overline{(M.N)}}.$$

---

[6]This is just the Cartesian product of M and N.

An aggregate with the cardinal number $a.b$ may also be made up out of two aggregates M and N with the cardinal numbers a and b according to the following rule: We start from the aggregate N and replace in it every element n by an aggregate $M_n \sim M$; if, then, we collect the elements of all these aggregates M_n to a whole S, we see that

(7) $$S \sim (M.N),$$

and consequently

$$\overline{\overline{S}} = a.b.$$

For, if, with any given law of correspondence of the two equivalent aggregates M and M_n, we denote by m the element of M which corresponds to the element m_n of M_n, we have

(8) $$S = \{m_n\};$$

thus the aggregates S and $(M.N)$ can be referred reciprocally and univocally to one another by regarding m_n and (m, n) as corresponding elements.

From our definitions result readily the theorems:

(9) $$a.b = b.a,$$

(10) $$a.(b.c) = (a.b).c,$$

(11) $$a.(b + c) = ab + ac;$$

because:

$$(M.N) \sim (N.M),$$
$$(M.(N.P)) \sim ((M.N).P),$$
$$(M.(N, P)) \sim ((M.N), (M.P)).$$

Addition and multiplication of powers are subject, therefore, to the commutative, associative, and distributive laws.

§4
The Exponentiation of Powers

By a "covering of the aggregate N with elements of the aggregate M," or, more simply, by a "covering of N with M," we understand a law by which with every element n of N a definite element of M is bound up, where one and the same element of M can come repeatedly into application. The element of M bound up with n is, in a way, a one-valued function of n, and may be denoted by $f(n)$; it is called a "covering function of n." The corresponding covering of N will be called $f(N)$.

Two coverings $f_1(N)$ and $f_2(N)$ are said to be equal if, and only if, for all elements n of N the equation

(1) $$f_1(n) = f_2(n)$$

is fulfilled, so that if this equation does not subsist for even a single element $n = n_0$, $f_1(N)$ and $f_2(N)$ are characterized as different coverings of N. For example, if m_0 is a particular element of M, we may fix that, for all n's

$$f(n) = m_0;$$

this law constitutes a particular covering of N with M. Another kind of covering results if m_0 and m_1 are two different particular elements of M and n_0 a particular element of N, from fixing that

$$f(n_0) \quad = m_0$$
$$f(n) \quad = m_1,$$

for all n's which are different from n_0.

The totality of different coverings of N with M forms a definite aggregate with the elements $f(N)$; we call it the "covering-aggregate of N with M" and denote it by $(N|M)$. Thus:

(2) $$(N|M) = \{f(N)\}.$$

If $M \sim M'$ and $N \sim N'$, we easily find that

(3) $$(N|M) \sim (N'|M').$$

Thus the cardinal number of $(N|M)$ depends only on the cardinal numbers $\overline{\overline{M}} = \mathfrak{a}$ and $\overline{\overline{N}} = \mathfrak{b}$; it serves us for the definition of $\mathfrak{a}^{\mathfrak{b}}$:

(4) $$\mathfrak{a}^{\mathfrak{b}} = \overline{\overline{(N|M)}}.$$

For any three aggregates, M, N, P, we easily prove the theorems:

(5) $$((N|M).(P|M)) \sim ((N, P)|M),$$

(6) $$((P|M).(P|N)) \sim (P|(M.N)),$$

(7) $$(P|(N|M)) \sim ((P.N)|M),$$

from which, if we put $\overline{\overline{P}} = \mathfrak{c}$, we have, by (4) and by paying attention to §3, the theorems for any three cardinal numbers, $\mathfrak{a}$, $\mathfrak{b}$, and $\mathfrak{c}$:

(8) $$\mathfrak{a}^{\mathfrak{b}}.\mathfrak{a}^{\mathfrak{c}} = \mathfrak{a}^{\mathfrak{b}+\mathfrak{c}},$$

(9) $$\mathfrak{a}^{\mathfrak{c}}.\mathfrak{b}^{\mathfrak{c}} = (\mathfrak{a}.\mathfrak{b})^{\mathfrak{c}},$$

(10) $$(\mathfrak{a}^{\mathfrak{b}})^{\mathfrak{c}} = \mathfrak{a}^{\mathfrak{b}.\mathfrak{c}}.$$

We see how pregnant and far-reaching these simple formulæ extended to powers are by the following example. If we denote the power of the linear continuum X (that is, the totality X of real numbers x such that $x \geq 0$ and ≤ 1) by $\mathfrak{c}$, we easily see that it may be represented by, amongst others, the formula:

(11) $$\mathfrak{c} = 2^{\aleph_0},$$

where §6 gives the meaning of $\aleph_0$. In fact, by (4), $2^{\aleph_0}$ is the power of all representations

(12) $$x = \frac{f(1)}{2} + \frac{f(2)}{2^2} + \cdots + \frac{f(\nu)}{2^\nu} + \cdots \quad \text{(where } f(\nu) = 0 \text{ or } 1\text{)}$$

of the numbers x in the binary system. If we pay attention to the fact that every number x is only represented once, with the exception of the numbers $x = \frac{2\nu+1}{2^\mu} < 1$, which are represented twice over, we have, if we denote the "enumerable" totality of the latter by $\{s_\nu\}$,

$$2^{\aleph_0} = \overline{\overline{(\{s_\nu\}, X)}}.$$

If we take away from X any "enumerable" aggregate $\{t_\nu\}$ and denote the remainder by X_1, we have:

$$X = (\{t_\nu\}, X_1) = (\{t_{2\nu-1}\}, \{t_{2\nu}\}, X_1),$$
$$(\{s_\nu\}, X) = (\{s_\nu\}, \{t_\nu\}, X_1),$$
$$\{t_{2\nu-1}\} \sim \{s_\nu\}, \quad \{t_{2\nu}\} \sim \{t_\nu\}, \quad X_1 \sim X_1;$$

so

$$X \sim (\{s_\nu\}, X),$$

and thus (§1)

$$2^{\aleph_0} = \overline{\overline{X}} = \mathfrak{c}.$$

From (11) follows by squaring (by §6, (6))

$$\mathfrak{c}.\mathfrak{c} = 2^{\aleph_0}.2^{\aleph_0} = 2^{\aleph_0+\aleph_0} = 2^{\aleph_0} = \mathfrak{c},$$

and hence, by continued multiplication by $\mathfrak{c}$,

(13) $$\mathfrak{c}^\nu = \mathfrak{c},$$

where ν is any finite cardinal number.

If we raise both sides of (11) to the power $\aleph_0$ we get

$$\mathfrak{c}^{\aleph_0} = (2^{\aleph_0})^{\aleph_0} = 2^{\aleph_0.\aleph_0}.$$

But since, by §6, (8), $\aleph_0.\aleph_0 = \aleph_0$, we have

(14) $$\mathfrak{c}^{\aleph_0} = \mathfrak{c},$$

The formulæ (13) and (14) mean that both the ν-dimensional and the $\aleph_0$-dimensional continuum have the power of the one-dimensional continuum. Thus the whole contents of my paper in Crelle's *Journal*, vol. lxxxiv, 1878, are derived purely algebraically with these few strokes of the pen from the fundamental formulæ of the calculation with cardinal numbers.

Cantor's idea to define exponentiation of cardinals by using the set of all functions from one set to another was a stroke of genius. Why this is a plausible generalization of exponentiation for finite numbers will be explored in the exercises.

The power of this idea can be seen in the example of the interval X of real numbers from 0 to 1. The numbers $f(v)$ in equation (12) are simply the zeros and ones in the binary expansion of the number x. Thus, the number x corresponds to a sequence of zeros and ones, in other words, a function f from the natural numbers $\{1, 2, \ldots, v, \ldots\}$ to the two-element set $\{0, 1\}$, and therefore the set of all such functions has, by Cantor's definition, the cardinality $2^{\aleph_0}$.

Cantor carefully notes that certain real numbers have two different representations in terms of sequences of zeros and ones. For instance, the number $\frac{1}{4}$ has the two binary representations $0.010000\ldots$ and $0.00111\ldots$. This set of real numbers with more than one binary expansion, he claims, is enumerable, in other words it is equivalent to the set of natural numbers. (The reader may show this as a challenging exercise or read Section 7 of Cantor's work.) A clever correspondence argument then shows that the double representations are not numerous enough to prevent X from having cardinality $2^{\aleph_0}$.

§6

The Smallest Transfinite Cardinal Number Aleph-Zero

Aggregates with finite cardinal numbers are called "finite aggregates," all others we will call "transfinite aggregates" and their cardinal numbers "transfinite cardinal numbers."

The first example of a transfinite aggregate is given by the totality of finite cardinal numbers v; we call its cardinal number (§1) "Aleph-zero" and denote it by $\aleph_0$; thus we define

(1) $$\aleph_0 = \overline{\overline{v}}.$$

That $\aleph_0$ is a *transfinite* number, that is to say, is not equal to any finite number μ, follows from the simple fact that, if to the aggregate $\{v\}$ is added a new element e_0, the union-aggregate $(\{v\}, e_0)$ is equivalent to the original aggregate $\{v\}$. For we can think of this reciprocally univocal correspondence between them: to the element e_0 of the first corresponds the element 1 of the second, and to the element v of the first corresponds the element $v + 1$ of the other. By §3 we thus have

(2) $$\aleph_0 + 1 = \aleph_0.$$

But we showed in §5 that $\mu + 1$ is always different from μ, and therefore $\aleph_0$ is not equal to any finite number μ.

The number $\aleph_0$ is greater than any finite number μ:

(3) $$\aleph_0 > \mu.$$

This follows, if we pay attention to §3, from the three facts that $\mu = \overline{\overline{(1, 2, 3, \ldots, \mu)}}$, that no part of the aggregate $(1, 2, 3, \ldots, \mu)$ is equivalent to the aggregate $\{\mu\}$, and that $(1, 2, 3, \ldots, \mu)$ is itself a part of $\{v\}$.

On the other hand, $\aleph_0$ is the least transfinite cardinal number. If $\mathfrak{a}$ is any transfinite cardinal number different from $\aleph_0$, then

(4) $$\aleph_0 < \mathfrak{a}.$$

This rests on the following theorems:

A. Every transfinite aggregate T has parts with the cardinal number $\aleph_0$.

Proof. If, by any rule, we have taken away a finite number of elements $t_1, t_2, \ldots, t_{\nu-1}$, there always remains the possibility of taking away a further element t_ν. The aggregate $\{t_\nu\}$, where ν denotes any finite cardinal number, is a part of T with the cardinal number $\aleph_0$, because $\{t_\nu\} \sim \{\nu\}$ (§1).

B. If S is a transfinite aggregate with the cardinal number $\aleph_0$, and S_1 is any transfinite part of S, then $\overline{\overline{S_1}} = \aleph_0$.

Proof. We have supposed that $S \sim \{\nu\}$. Choose a definite law of correspondence between these two aggregates, and, with this law, denote by s_ν that element of S which corresponds to the element ν of $\{\nu\}$, so that

$$S = \{s_\nu\}.$$

The part S_1 of S consists of certain elements s_κ of S, and the totality of numbers κ forms a transfinite part K of the aggregate $\{\nu\}$. By Theorem G of §5 the aggregate K can be brought into the form of a series

$$K = \{\kappa_\nu\},$$

where

$$\kappa_\nu < \kappa_{\nu+1};$$

consequently we have

$$S_1 = \{s_{\kappa_\nu}\}.$$

Hence it follows that $S_1 \sim S$, and therefore $\overline{\overline{S_1}} = \aleph_0$.

From A and B the formula (4) results, if we have regard to §2.

From (2) we conclude, by adding 1 to both sides,

$$\aleph_0 + 2 = \aleph_0 + 1 = \aleph_0,$$

and, by repeating this

(5)
$$\aleph_0 + \nu = \aleph_0.$$

We also have

$$\aleph_0 . \aleph_0 = \aleph_0.$$

Proof. By (6) of §3, $\aleph_0 . \aleph_0$ is the cardinal number of the aggregate of bindings

$$\{(\mu, \nu)\},$$

where μ and ν are any finite cardinal numbers which are independent of one another. If also λ represents any finite cardinal number, so that $\{\lambda\}$, $\{\mu\}$, and $\{\nu\}$ are only different notations for the same aggregate of all finite numbers, we have to show that

$$\{(\mu, \nu)\} \sim \{\lambda\}.$$

Let us denote $\mu + \nu$ by ρ; then ρ takes all the numerical values $2, 3, 4, \ldots$, and there are in all $\rho - 1$ elements (μ, ν) for which $\mu + \nu = \rho$, namely:

$$(1, \rho - 1), (2, \rho - 2), \ldots, (\rho - 1, 1).$$

In this sequence imagine first the element $(1, 1)$, for which $\rho = 2$, put, then the two elements for which $\rho = 3$, then the three elements for which $\rho = 4$, and so on. Thus we get all the elements (μ, ν) in a simple series:

$$(1, 1); (1, 2), (2, 1); (1, 3), (2, 2), (3, 1); (1, 4), (2, 3), \ldots,$$

and here, as we easily see, the element (μ, ν) comes at the λth place, where

(9) $$\lambda = \mu + \frac{(\mu + \nu - 1)(\mu + \nu - 2)}{2}.$$

The variable λ takes every numerical value $1, 2, 3, \ldots$, once. Consequently, by means of (9), a reciprocally univocal relation subsists between the aggregates $\{\lambda\}$ and $\{(\mu, \nu)\}$.

Recall from Section 4 that the real numbers between zero and one have cardinality $2^{\aleph_0}$. In the 1891 paper *Über eine Elementare Frage der Mannigfaltigkeitslehre* (On an Elementary Question in the Theory of Sets) [28], from which we include an excerpt below [28, pp. 278 f.], Cantor shows that if a is a cardinal number, then $2^a > a$. In particular, $2^{\aleph_0} > \aleph_0$. This means that the set of real numbers has larger cardinality than the natural (and rational) numbers.

Georg Cantor, from
On an Elementary Question in the Theory of Sets

Namely, if m and w are any two distinct characters, we form a collection M of elements

$$E = (x_1, x_2, \ldots, x_\nu, \ldots)$$

which depends on infinitely many coordinates $x_1, x_2, \ldots, x_\nu, \ldots$, each of which is either m or w. Let M be the set of all elements E.

Amongst the elements of M are for example the following three

$$E^{\mathrm{I}} = (m, m, m, m, \ldots),$$
$$E^{\mathrm{II}} = (w, w, w, w, \ldots),$$
$$E^{\mathrm{III}} = (m, w, m, w, \ldots).$$

I now claim that such a manifold[7] M does not have the power of the series $1, 2, \ldots, \nu, \ldots$.

---

[7] Set.

This follows from the following theorem:

"If $E_1, E_2, \ldots, E_\nu, \ldots$ is any simply infinite sequence of elements of the manifold M, then there is always an element E_0 of M which does not agree with any E_ν."

To prove this let

$$E_1 = (a_{11}, a_{12}, \ldots, a_{1\nu}, \ldots),$$
$$E_2 = (a_{21}, a_{22}, \ldots, a_{2\nu}, \ldots),$$
$$\ldots$$
$$E_\mu = \left(a_{\mu 1}, a_{\mu 2}, \ldots, a_{\mu \nu}, \ldots\right).$$
$$\ldots$$

Here each $a_{\mu\nu}$ is a definite m or w. A sequence $b_1, b_2, \ldots, b_\nu$ will now be so defined, that each b_ν is also equal to m or w and *different* from $a_{\nu\nu}$.

So if $a_{\nu\nu} = m$, then $b_\nu = w$, and if $a_{\nu\nu} = w$, then $b_\nu = m$.

If we then consider the element

$$E_0 = (b_1, b_2, b_3, \ldots)$$

of M, then one sees immediately that the equation

$$E_0 = E_\mu$$

cannot be satisfied for any integer value of μ, since otherwise for that value of μ and all integer values of ν

$$b_\nu = a_{\mu\nu}$$

and also in particular

$$b_\mu = a_{\mu\mu},$$

which is excluded by the definition of b_μ. It follows immediately from this theorem that the totality of all elements of M cannot be put in the form of a series: $E_1, E_2, \ldots, E_\nu, \ldots$, since otherwise we would have the contradiction that a thing E_0 would be an element of M and also not an element of M.

This proof appears remarkable not only due to its great simplicity, but in particular for the reason that the principle employed in it can be directly extended to the general theorem, that the powers of well-defined point sets have no maximum, or what amounts to the same, that to every given point-set L can be associated another one M which has a higher power than L.

Exercise 2.14: For $M = \{1, 2, 3, 4, 5\}$ and $N = \{a, b, c, d, e, f, g, h\}$, what are $\overline{\overline{M}}$ and $\overline{\overline{N}}$? Show that $\overline{\overline{M}} < \overline{\overline{N}}$ using the definition given in §2.

Exercise 2.15: Verify the distributive law for powers; that is, prove (11) in §3.

Exercise 2.16: If A, B, C are sets, show that $A.(B.C) \neq (A.B).C$, but they are equivalent.

Exercise 2.17: Which real numbers between 0 and 1 have more than one binary expansion? How many different binary expansions for a number are possible?

Exercise 2.18: Why are covering functions so named?

Exercise 2.19: What are 0^0, 0^1, and 1^0?

Exercise 2.20: Prove (3) in §4.

Exercise 2.21:

1. Let M and N be aggregrates such that $\overline{\overline{M}} = 3$ and $\overline{\overline{N}} = 2$. Show that $3^2 = \overline{\overline{(N \mid M)}}$.

2. If $\overline{\overline{M}} = m$ and $\overline{\overline{N}} = n$ where m and n are positive integers, show that $m^n = \overline{\overline{(N \mid M)}}$.

3. Suppose $\overline{\overline{M}} = m$ where m is a positive integer, and let $\mathcal{P}(M)$ denote the power set of M (i.e., the set of all subsets of M). Show that

$$\overline{\overline{\mathcal{P}(M)}} = \overline{\overline{(M \mid 2)}} = 2^m.$$

Exercise 2.22: Prove (8), (9), and (10) in §4.

Exercise 2.23: Show that $\aleph_0 + \aleph_0 = \aleph_0$.

Exercise 2.24: Verify formula (9) of §6 by representing Cantor's series of ordered pairs (or "bindings" as he calls them)

$$(1, 1); (1, 2), (2, 1); (1, 3), \ldots$$

as points in the plane and indicating their order of precedence.

Exercise 2.25: Generalize Cantor's argument in the last source to prove that $2^m > m$ for any cardinal number m.

Exercise 2.26: List an infinite sequence of infinite cardinal numbers.

Exercise 2.27: Show that Frege's proposal for a definition of cardinal numbers leads to a paradox. Recall that he proposes to define the number 1, for instance, as the equivalence class of all sets with one element, that is,

$$1 = \{\{a\} \mid a \text{ is a set}\}.$$

Exercise 2.28: Read Cantor's §7 on ordinal types.

2.4 Zermelo's Axiomatization

Ernst Zermelo's work is situated at two opposite ends of the mathematical spectrum. He wrote his dissertation in 1899 at the University of Göttingen on the very applied topic "Hydrodynamic Investigations of Currents on the Surface of a Sphere." Five years later he proved his Well-Ordering Theorem, far removed from

PHOTO 2.5. Zermelo.

any mathematics that seemed applicable at the time. Almost all of his publications fall into either the area of physics and the calculus of variations, or are related to set theory. A notable combination of applied and pure mathematics can be found in a paper he wrote on applications of set theory to chess. He is best known through his axiomatization of Cantorian set theory, which represents his most important contribution.

Following are excerpts (pp. 261–267) from Zermelo's 1908 paper *Untersuchungen über die Grundlagen der Mengenlehre, I* (Investigations on the Foundations of Set Theory, I) [180], in which he proposed his axiomatization of Cantorian set theory. As mentioned in the introduction, Zermelo was somewhat naive about the question of consistency and independence of his axioms, as future developments made amply clear.

Ernst Zermelo, from

Investigations on the Foundations of Set Theory, I

Set theory is that part of mathematics whose task it is to investigate mathematically the basic concepts of number, order, and function, in the simplicity of their origins, and thereby to develop the logical foundations of all of arithmetic and analysis; thus forming an indispensable part of mathematical science. It now appears, however, that just this discipline is threatened in its very existence by certain contradictions, or "antinomies," which apparently can be derived from its necessary principles of thought, and which have not yet found an entirely satisfactory solution. Especially in light of "Russell's antinomy" of the "set of all sets which do not contain themselves as an element",[*] it does not seem to be admissible today to assign to an arbitrary logically definable concept a "set" or "class" as its "size." Cantor's original definition of a "set" as a "collection into a whole of definite and separate objects of our intuition or our thought"[†] thus in any case is in need of a restriction. The effort to replace it by a different equally simple one, which would not give rise to such concerns, has not yet been successful. Under these circumstances, the only recourse left at this time is to take the converse route and, beginning with "set theory" as it historically exists, to find the principles which are required for the foundation of this mathematical discipline. This problem has to be solved in such a way that on the one hand one makes the principles sufficiently narrow to exclude all contradictions, but at the same time makes them sufficiently broad to retain everything that is valuable in this science.

In the present paper I plan to show how the complete theory created by G. Cantor and R. Dedekind can be reduced to a handful of definitions and seven "principles" or "axioms" which are apparently independent of each other. Here we leave aside the further, more philosophical, question about the origin and domain of validity of these "principles." I have not even been able to give a rigorous proof for the "consistency" of my axioms, which is certainly very essential. Instead, I had to limit myself to the occasional remark that all the "antinomies" known up to now disappear if one uses the principles suggested here as a foundation. With this work I would like at least to provide some useful preparation for later investigations of deeper problems of this kind.

The following paper contains the axioms and their immediate consequences, as well as a theory of equivalence based on these principles, which avoids the formal use of cardinal numbers. A second article, which will develop the theory of well-ordering and its application to finite sets and the principles of arithmetic, is in preparation.

§1.
Basic Definitions and Axioms.

1. Set theory concerns itself with a "domain" $\mathfrak{B}$ of objects which we simply want to call "things," and of which the "sets" form a part. If two symbols a and b indicate the same thing, then we write $a = b$, in the opposite case, $a \neq b$. We say of a thing a that it "exists" if it belongs to the domain $\mathfrak{B}$; [. . .]

2. Between the things of the domain $\mathfrak{B}$ there exist certain *basic relationships* of the form $a \in b$. If for two things a, b the relationship $a \in b$ holds, we say that "a is

---

[*]B. Russell, "The Principles of Mathematics," pp. 366–368, 101–107.
[†]G. Cantor, *Math. Annalen* Bd. 46, p. 481.

an *element* of the set b," or "b contains a as an element," or "contains the element a." A thing b which contains another a as element can always be called a *set*, but only then — with a single exception (Axiom II).

3. If every element x of a set M is at the same time also an element of the set N, so that from $x \in M$ we can always conclude $x \in N$, then we say "M is a *subset* of N," and write $M \subseteq N$.

· · ·

Now the following "*axioms*" or "*postulates*" hold for the basic relationships of our domain $\mathfrak{B}$.

Axiom I. If every element of a set M is at the same time an element of N, and conversely, that is if at the same time $M \subseteq N$ and $N \subseteq M$, then we always have $M = N$. Or, more briefly: every set is determined by its elements.

(Axiom of Determinacy.)

· · ·

Axiom II. There is an (improper) set, the "*null set*" 0, which does not contain any elements at all. If a is any thing of the domain, then there exists a set $\{a\}$ which contains a and only a as an element; if a and b are any two things of the domain, then there always exists a set $\{a, b\}$ which contains both a and b as elements, but no thing different from those two.

(Axiom of Elementary Sets.)

· · ·

[Zermelo calls a property $\mathfrak{F}(x)$ "definite" for all elements x of a set M if the basic relationships of $\mathfrak{B}$ and the laws of logic determine whether $\mathfrak{F}(x)$ holds for each element x of M.]

Axiom III. If the [property] $\mathfrak{F}(x)$ is definite for all elements of a set M, then M always contains a subset $M_{\mathfrak{F}}$ which contains all those elements x of M for which $\mathfrak{F}(x)$ is true, and only those.

(Axiom of Separation.)

· · ·

Axiom IV. To every set T there corresponds a second set $\mathfrak{U}T$ (the "*power set*" of T),[8] which contains all subsets of T as elements, and only those.

(Power Set Axiom.)

· · ·

---

[8]The letter U ($\mathfrak{U}$) was presumably chosen by Zermelo because it is the first letter of the German word for "subset," *Untermenge*.

Axiom V. To every set T there corresponds a set $\mho T$ (the "*Union Set*" of T, which contains all elements of the elements of T as elements, and only those.[9]

(Axiom of Union.)

$\cdots$

Axiom VI. If T is a set, all of whose elements are sets which are different from 0 and are pairwise disjoint, then their union $\mho T$ contains at least one subset S_1 which has exactly one element in common with every element of T.

(Axiom of Choice.)

$\cdots$

Axiom VII. The domain contains at least one set Z which contains the null set as element, and is such that to each of its elements a there corresponds another element of the form $\{a\}$ [...].

(Axiom of Infinity.)

Zermelo's axioms postulate the existence of two sets; the empty set and an infinite set. All other sets that can be constructed within this framework are built from existing ones through certain narrowly specified tools, such as the formation of subsets, power sets, the axiom of choice, and union. Clearly, the "set of all sets" or the "set of all sets that do not contain themselves as an element" are far beyond the reach of this modest collection of tools.

In 1922, it was shown by Adolf Fraenkel [61] that Zermelo's axioms were not quite enough to construct all sets in Cantor's theory. Zermelo's Axiom VII postulates the existence of an infinite set $Z = Z_0$. Let Z_1 denote the power set of Z_0, Z_2 the power set of Z_1, etc. Fraenkel pointed out that Zermelo's axioms were not strong enough to allow the construction of the set $\{Z_0, Z_1, Z_2, \ldots\}$ (Exercise 2.33), and hence one could also not construct the union of all the sets in this set. He then observes that if one assigns a cardinal number $< \aleph_\omega$ to the continuum, then one cannot prove the existence of sets with cardinality greater than or equal to $\aleph_\omega$. As a remedy, Fraenkel suggested the addition of the following axiom:

Substitution Axiom: If M is a set, and if every element of M is replaced by an object from the domain $\mathfrak{B}$ (see Zermelo's paper), then another set results.

Fraenkel suggested a few more minor modifications, and the resulting axiom system is now known as Zermelo–Fraenkel set theory, finally providing the standard foundation for modern mathematics.

Exercise 2.29: What are the "primitive" undefined terms in Zermelo's system?

Exercise 2.30: Derive the first part of Axiom II from Axiom III.

---

[9]The letter G ($\mho$) is the first letter of the German word for "totality," *Gesamtheit*.

Exercise 2.31: Let $T = \{a, b, \{x, y\}\}$. Find $\mathfrak{U}T$, $\mathfrak{G}T$, $\mathfrak{U}(\mathfrak{G}T)$, and $\mathfrak{G}(\mathfrak{U}T)$.

Exercise 2.32: Use Zermelo's axioms to construct a set with

1. exactly three elements.
2. exactly n elements, for every positive integer n.
3. countably many elements.

Exercise 2.33: Show that Zermelo's axiom system does not allow the construction of the set $\{Z_0, Z_1, \ldots\}$ above.

Exercise 2.34: There are still controversies today about the foundations of mathematics. Read about intuitionism, constructivism, and formalism. (See, e.g., [5].)

CHAPTER 3

Analysis: Calculating Areas and Volumes

3.1 Introduction

In 216 B.C.E., the Sicilian city of Syracuse made the mistake of allying itself with Carthage during the second Punic war, and thus was attacked by Rome, portending what would ultimately happen to the entire classical Greek world. During a long siege, soldiers of the Roman general Marcellus were terrified by ingenious war machines defending the city, invented by the Syracusan Archimedes, greatest mathematician of the ancient world, born in 287 B.C.E. These included catapults to hurl great stones, as well as ropes, pulleys, and hooks to raise and smash Marcellus's ships, and perhaps even burning mirrors setting fire to their sails. Finally, though, probably through betrayal, Roman soldiers entered the city in 212 B.C.E., with orders from Marcellus to capture Archimedes alive. Plutarch relates that "as fate would have it, he was intent on working out some problem with a diagram and, having fixed his mind and his eyes alike on his investigation, he never noticed the incursion of the Romans nor the capture of the city. And when a soldier came up to him suddenly and bade him follow to Marcellus, he refused to do so until he had worked out his problem to a demonstration; whereat the soldier was so enraged that he drew his sword and slew him" [93, p. 97]. Despite the great success of Archimedes' military engineering inventions, Plutarch says that "He would not deign to leave behind him any commentary or writing on such subjects; but, repudiating as sordid and ignoble the whole trade of engineering, and every sort of art that lends itself to mere use and profit, he placed his whole affection and ambition in those purer speculations where there can be no reference to the vulgar needs of life" [93, p. 100]. Perhaps the best indication of what Archimedes truly loved most is his request that his tombstone include a cylinder circumscribing a sphere, accompanied by the inscription of his amazing theorem that the sphere is exactly two-thirds of the circumscribing cylinder in both surface area and volume!

PHOTO 3.1. The death of Archimedes.

Contrast this with what one typically learns about areas and volumes in school. One is told that the area inside a circle is πr^2, with π itself remaining mysterious, that the volume of a sphere is $\frac{4}{3}\pi r^3$, and perhaps that its surface area is $4\pi r^2$. (The reader can easily convert this information into Archimedes' grave inscription.) Are these formulas, all discovered and proved by Archimedes in the aesthetically pleasing form stated at his grave, perhaps quite literally the only areas or volumes the reader "knows" for regions with curved sides? If so, is this because these particular geometric objects are incredibly special, while the precise areas and volumes of all other curved regions remain out of human reach, their curving sides making exact calculations unattainable?

The two-thousand-year quest for areas and volumes has yielded many strikingly beautiful results, and is also an extraordinary story of discovery. This challenge led to the development of integral calculus, a systematic method for finding areas and volumes discovered in the seventeenth century. (Excellent references for this development are [8, 21].) Calculus became the driving force of eighteenth-century mathematics, with far-reaching applications, especially involving motion and other types of change. Subsequent efforts to understand, justify, and expand its applicability spurred the modern mathematical subject known as "analysis" [20, 77, 83, 93, 97]. Our sources will trace the development of the calculus through the problem of finding areas. And we shall encounter philosophical controversies that boiled at the heart of mathematics for two millennia, with a dramatic turn of events as late as 1960.

Cultures prior to classical Greece amassed much knowledge of spatial relations, including some understanding of areas and volumes, motivated by endeavors such as resurveying and reshaping fields after floods, measuring the volume of piles of grain for taxation or community food planning, or determining the amounts of needed materials for constructing pyramids, temples, and palaces. Their knowledge resulted largely from empirical investigations, or inductive generalization by analogy, from simpler to more complicated situations, without rigorous deductive proof [91]. Babylonian and Egyptian mathematics was greatly restricted by mostly considering concrete cases with definite numbers, rather than general abstract objects [91]. It was the emerging mathematical culture of ancient Greece that embraced abstract concepts in geometry, and combined them with logic to implement deductive methods of proof with great power to discover and be certain of the truth of new results [84, 85]. Probably introduced by Thales around 600 B.C.E., these methods were firmly established by Pythagoras and his school during the sixth century.

The Pythagoreans' discovery that the diagonal of a square is incommensurable with its side greatly influenced the course of classical Greek mathematics. What they found is that no unit, however small, can be used to measure both these lengths (hence the term incommensurable; today we would say that their ratio is "irrational"). And yet to Greek mathematicians it was the very essence of a number to be expressed as some multiple of a chosen unit (i.e., a natural number). While they were able to use units to work with ratios comparing natural numbers (i.e., fractions), the discovery that this sort of comparison was impossible in general for geometric lengths caused them to reject numbers alone as sufficient for measuring the magnitudes of geometry. The study of areas was therefore not pursued by assigning a number to represent the size of each area, but rather by directly comparing areas, as on Archimedes' gravestone.

Perhaps the first solution to an area problem for a region with curved edges was Hippocrates of Chios's theorem in the mid–fifth century B.C.E., about a certain lune, the crescent-moon-shaped region enclosed by two intersecting circles (Figure 3.1) [45, pp. 17 f.]. Here semicircles and their diameters are arranged as shown, with CD perpendicular to the large diameter AB at its center. Hippocrates showed that the lune $AECFA$ equals ("in area," we would say) triangle ACD (Exercise 3.1). To accomplish this, he needed the theorem that the ratio (of areas) of two circles is the ratio of the squares of their diameters, which was part of a large body of deductive knowledge already familiar to him [93, Ch. 2].

Hippocrates' result is called a "quadrature," or "squaring," of the lune, since the idea was to understand the magnitude of an area by equating it to that of a square of known size, i.e., one whose side was a "known" magnitude. A known quantity was one that one could construct using compass and straightedge alone, or equivalently, with the postulates of Euclid's *Elements* (see the geometry chapter). In practice, areas were often effectively squared by equating them to known triangles and rectangles, since with straightedge and compass one can always easily "square" the resulting triangle or rectangle [45, pp. 13 f.][84, 85].

Hippocrates also tried to "square the circle," i.e., construct a square with area precisely that of a given circle. Two other prominent construction problems from

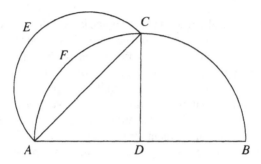

FIGURE 3.1. A lune squared by Hippocrates

this era were "doubling the cube" (see the Introduction to the algebra chapter) and "trisecting the angle," i.e., constructing an angle one-third the size of a given angle. These are called the three classical problems of antiquity, and they motivated much of Greek mathematics for centuries, including investigations into what modifications of the construction rules might be necessary for solving them [98]. It is astonishing that all three took more than two thousand years to be solved, and the answers are all the same: each of the constructions is literally impossible with compass and straightedge alone (Exercises 3.2 and 3.3).

In the fifth century B.C.E. Democritus discovered that the volume of a cone is precisely one-third that of the encompassing cylinder with same base and height. We not know how he derived these results, but we do know that he considered solids as possibly "made up" of an infinite number of infinitely thin layers, infinitely near together, for he said:

> If a cone were cut by a plane parallel to the base [and very close to the base], what must we think of the [areas of the] surfaces forming the sections? Are they equal or unequal? For, if they are unequal, they will make the cone irregular as having many indentations, like steps, and unevenness; but if they are equal, the sections will be equal, and the cone will appear to have the property of the cylinder, and to be made up of equal, not unequal, circles: which is very absurd [85, p. 119].

This paradox did not prevent exploitation of the idea that a solid is made up of these infinitely thin slices, or that a planar region such as a disk is made up of infinitely many infinitely thin parallel line segments. But the mathematical and philosophical foundations for such an approach would remain murky and controversial for two millennia. These infinitely thin objects would later be called "indivisibles" because, having no thickness, they cannot be divided. A certain schizophrenia ruled between the largely successful, but rigorously unsupported, use of indivisibles for calculating areas and volumes, and fully accepted rigorous deductive methods that were much more difficult to apply.

Related philosophical trends fueled the controversy. Plato distinguished between the objects of the mind and those of the physical world, e.g., an ideal mathematical circle versus a physical circle. In the context of the Democritean puzzle, this provoked questions about a distinction between indivisible physical and mathematical objects. Such questions were amplified when space and time were combined in the four paradoxes of Zeno of Elea (c. 450 B.C.E.). One of his paradoxes, the "arrow" [Aristotle, *Physics* VI, 9, 239^{b}6], refutes the existence of indivisibles. It claims that if an instant is indivisible, an arrow in flight cannot move during the instant, since otherwise the instant could be divided. But since time is made up only of such instants, the arrow can therefore never move.

A satisfactory mathematical response to this paradox did not gel until the nineteenth century. Classical Greek work sidestepped the dilemma by an incredibly ingenious approach avoiding indivisibles. Today called the "method of exhaustion," it was invented by Eudoxus in the fourth century B.C.E. to prove Democritus's discoveries on volumes of cones and pyramids. He proved results for curved figures by providing geometric approximations (usually with flat sides) whose area or volume was already known. If such approximations were found that together "exhaust" (i.e., fill out) the curved figure, then the exact result could often be verified. The method of exhaustion was featured around 300 B.C.E. in Book XII of the *Elements*, Euclid's compilation of much of Greek mathematics. Here Euclid proved the earlier results of Hippocrates and Democritus, as well as others, for instance that the ratio of the volumes of two spheres is the ratio of the cubes of their diameters (check this using the modern formula for the volume of a sphere). Euclid taught and wrote at the Museum and Library (something like a university) founded around that time in Alexandria by Ptolemy I Soter. It was the intellectual focal point of Greek scholarship for centuries, and eventually contained over 500,000 texts.

The exhaustion method was elevated to an art form by the great mathematician Archimedes of the third century B.C.E. He proved a comprehensive array of area and volume results, without equal until at least the seventeenth century. Our first original sources will come from two treatises of Archimedes on the area of a segment of a parabola. They reveal that while he felt obliged to provide proofs by the rigorous method of exhaustion, exemplified in *Quadrature of the Parabola*, his actual discovery tools, revealed in *The Method*, were full of the fruitful indivisibles excluded from formal proofs.

After Archimedes, several factors conspired to stymie further calculation of areas and volumes. One was the inability of Greek mathematics and philosophy to resolve the paradoxes attached to indivisibles, and associated issues about infinity and the nature of space and time. Second, Archimedes' *Method* was lost for two thousand years, requiring the rediscovery of its ideas by later generations. Creativity was also hampered by the meager interest the Roman conquerors of Greece showed in mathematics, science, or philosophy; their focus was on practical arts and engineering projects. Finally, it was difficult to press further with the method of exhaustion simply because it was cumbersome and not well suited to discovering new results.

The mathematics of classical Greece declined over several centuries, and by the fourth century C.E., Greek mathematical activity had effectively ceased. This end is marked by the barbaric murder in Alexandria of Hypatia, the first female mathematician we know much about, by a Christian mob [93]. The emergence of algebra in the Arabic world, with influences from Greek and Indian traditions, set the stage for later developments. After the founding of Islam by Mohammed (570–632), his followers made many military conquests, including Alexandria by 641, where they burned what books still remained in the great Library there. Yet within another century, a great cultural awakening in Islam established the "House of Wisdom" in Baghdad, comparable to the Museum and Library in Alexandria, with avid translation of Greek texts into Arabic (see the Introduction to the algebra chapter). One of the most important influences there was the mathematician Al-Khwarizmi, who in the early ninth century began to develop algebra, a general set of procedures for solving certain problems, and also wrote on the Hindu decimal numeration system with place values and zero [20, pp. 250 f.]. These advances made it possible to start generalizing arithmetic processes using algebraic symbols, and thus to calculate in more general situations [8, pp. 60 f.].

For many centuries Europe had little access to classical Greek mathematics, cut off by Arabic conquests. But in twelfth-century Spain and Sicily, Arabic, Greek, and Hebrew texts were translated into Latin and began to filter into Europe. Classical Greek mathematics, Arabic algebra, and the Hindu numeral system in Arabic texts began to affect European thinkers. In particular, the writings of the great classical Greek scholar Aristotle (384–322 B.C.E.) spurred lively argument among thirteenth and fourteenth century philosophers about the infinite, indivisibles, and continuity [8, 21].

An illustrative example of these discussions concerns two concentric circles. Each radial line cuts the circumference of each circle in exactly one point, setting up a one-to-one correspondence between the points on the larger circumference and those on the smaller. This suggests that the number of points on each circumference is exactly the same, and yet there would seem to be more on the outer one, since it is longer. The paradox that the longer curve has all its points in perfect correspondence with those of the shorter one was unresolvable to fourteenth-century thinkers, and would remain a puzzle for several more centuries, until mathematicians began to understand indivisibles. (See the set theory chapter for more on this.)

Medieval scholars also began a quantitative study of physical change, in particular the effect that continuous variation has on the ultimate form of something [8, pp. 81 f.][21]. The reader can see the connection to Democritus's ancient quotation on the volume of a cone: Each indivisible slice is a circular disk, but they vary continuously in area as we slide along the cone, ultimately combining to produce the volume of the cone. Exactly how quickly the disks vary in area will be crucial to measuring the total volume. Medieval study of how variation influences form is called the "latitude of forms"; while it was primarily philosophical, it opened the door to later mathematical work, and pioneered perpendicular coordinate axes as a way of representing change such as velocity and acceleration.

Aristotle's view, which heavily influenced early medieval thinkers, was dominated by a strong reliance upon sensory perception, and a reluctance to abstract and extrapolate beyond what was clearly derived from the physical world. He had argued that the process of successively dividing a magnitude could always continue further, and thus indivisibles could not exist. He also argued that because a number can always be increased, but the world is finite, the infinitely large is only potential, and not actual: "In point of fact they [mathematicians] do not need the infinite and do not use it. They postulate only that the finite straight line may be produced as far as they wish.... Hence, for the purposes of proof, it will make no difference to them to have such an infinite instead..." [20, p. 41].

However, a fifteenth-century trend toward Platonic and Pythagorean mysticism, accompanying the growth of humanism over the extremely rational and rigorous thought of earlier Scholastic philosophy, encouraged the previously denied use of the infinite and infinitesimal in geometry. As a result of this shift, mathematics was viewed as independent of the senses, not bound by empirical investigations, and thus free to use the infinite and infinitesimal, provided that no inconsistencies resulted. By the early seventeenth century this view opened up bold new approaches.

An early trailblazer was Johannes Kepler (1571–1630), a German astronomer and mathematician [8, pp. 108 f.][21, pp. 106 f.][77, pp. 11 f.]. He deduced his three famous laws of planetary motion from the incredibly detailed astronomical observations made by Danish astronomer Tycho Brahe prior to the availability of the telescope. Kepler's work supported Copernicus's heliocentric model for the solar system, demonstrating that the planets follow simple elliptical orbits around the sun. His laws describe these motions, including the relationship between the period of each orbit and the size of its ellipse. One of the greatest triumphs of the newly invented calculus later in the century was Isaac Newton's (1642–1727) mathematical derivation of Kepler's laws from Newton's own physical theories of force and gravitation.

Kepler made free use of indivisibles in both astronomical work and a treatise on measuring volumes of wine casks. He went far beyond the practical needs of the wine business, and wrote an extensive tract on indivisible methods. Two illustrative examples are his approaches to the areas of a circle and an ellipse (Exercise 3.4) [20, pp. 356 f.].

One of Kepler's contemporaries was the great Italian mathematician, astronomer, and physicist Galileo Galilei (1564–1642), the first to use a telescope to study the heavens, and often considered the founder of the experimental method and modern physics. A university professor at Pisa and Padua, Galileo deduced the laws of freely falling bodies and the parabolic paths of projectiles, initiating an era of applications of mathematics to physics. In his book *Two New Sciences*, he used indivisible methods to study the motion of a falling body, and he planned, but never wrote, an entire book on indivisibles [77, pp. 11–12],[93]. Galileo established mathematical rationalism against Aristotle's approach to studying the universe, and insisted that "The book of Nature is...written in mathematical characters" [49, vol. 19, p. 640]. His support of the Copernican theory that the earth orbits the sun caused his prosecution by the Roman Catholic Inquisition, forcible recantation of his ideas, and house arrest for the last eight years of his life. In 1979, Pope John Paul II appointed Vatican specialists to study the case, and after thirteen more years, in a 1992 address to the Pontifical Academy of Sciences, the Pope officially claimed the persecution of Galileo had resulted from a "tragic case of mutual incomprehension" [57]. Thus it took 350 years for the church to "rehabilitate" Galileo, and *Two New Sciences* remained on the church Index of forbidden books until 1822.

We will examine a text using indivisibles by Bonaventura Cavalieri (1598–1647), a pupil and associate of Galileo who also became a university professor. Cavalieri combined indivisibles with emerging algebraic techniques to produce many new insights into the problem of finding areas and volumes.[1] Archimedes had determined the area of a parabolic segment, which we will see is equivalent to finding the area bounded above by a portion of the curve $py = x^2$ (p a constant) and below by the x-axis. Cavalieri managed to generalize this to calculate areas bounded by "higher" parabolas $py = x^n$, although he had neither this modern notation nor our modern view of the curve as a functional relationship between two variables.

The explosion of techniques using indivisibles in the early to mid–seventeenth century is a good example of simultaneous independent discovery, leading to disputes about plagiarism and priority, exacerbated by mathematicians' reluctance to reveal their methods. Publishing was also difficult, since the first scientific periodicals came into existence only in the latter third of the century [77]. Much of the exchange of information occurred through correspondence with one man, Marin Mersenne, in Paris, who circulated the problems and manuscripts of others. Alternative approaches to higher parabolas and other curves and surfaces via indivisibles came from Cavalieri's fellow Italian Evangelista Torricelli (1608–1647), Frenchmen Pierre de Fermat (1601–1665), Blaise Pascal (1623–1662), Gilles Personne de Roberval (1602–1675), and Englishman John Wallis (1616–1703). Some of them were simultaneously pursuing the important "problem of tangents," which we will see united in the most spectacular way with the problem of finding areas.

Although certain curves defined by simple geometric relationships were inherited from the Greeks, the repertoire had been quite limited. Now there was a great expansion in the curves considered, such as the "cycloid" (the path traveled by a point on the edge of a rolling wheel), higher parabolas and hyperbolas, and the "catenary" formed by the shape of a hanging chain. There was an interplay with physics and optics, great interest in lengths of curves, and in properties of curves arising from mechanical motion, like that of a pendulum. Their analysis required a melding of geometry and algebra, overcoming the classical aversion to linking geometric measurements to numbers and to the algebraic equations that arose from studying numbers [77].

Francois Viète (1540–1603) had contributed much to the introduction of symbols into the previously largely verbal art of equations, making it easier to define and work with equations corresponding to geometrical constructions, and thus be-

---

[1] A method similar to Cavalieri's had already been used in China beginning in the third century to find the volume of a sphere, by comparing it to the "double box lid" obtained by intersecting two perpendicular cylinders inside a cube [116, pp. 282 ff.][165].

ginning to reconnect geometry with algebra. This direction expanded greatly in work of Fermat and René Descartes (1596–1650) (after whom Cartesian coordinates are named). Descartes worked on finding the line perpendicular to a curve at any point, and others such as Roberval and Fermat worked on the equivalent problem of finding the tangent to a curve at any point. To them this tangent line embodied the direction of motion of a point moving along the curve, or was interpreted as the line touching the curve at only that one point. Their ingenious methods were highly successful, but involved distances or magnitudes becoming vanishingly small, and thus controversy about their meaning and validity. Another problem studied with similar methods was that of "maxima and minima," i.e., finding the extreme values of a quantity or, in terms of curves, finding the highest or lowest point. Altogether the seventeenth century was an incredibly rich time of ferment and invention. (See also the Introduction to the number theory chapter.)

While there were indications in some of the new area and volume calculations that they were connected to the results of tangent problems, it was the next generation, specifically Isaac Newton, in England, and Gottfried Leibniz (1646–1716), in France and Germany, who transformed these indications into an explicit connection and exploited it as a tool for a tremendous leap. Working in the latter part of the seventeenth century, Newton and Leibniz not only explicitly recognized a connection between area and tangent problems that provided a general method for solving area, volume, and motion problems, but they also systematically explored the translation of such problems into formulaic guise, leading to a greater understanding of these phenomena in a more general and calculational setting. They invented a symbolic language to express and exploit these connections; thereby a new mathematical subject, the calculus, was born.

The beautiful relationship between areas and tangents will emerge as we read Leibniz's proof of the Fundamental Theorem of Calculus, providing his general way of solving area problems. The process involved in finding an area is today known as *integration* (or calculating an *integral*, as Leibniz himself called it). Using modern function terminology, if the graph of $y = f(x)$ is a curve lying above the x-axis between the vertical lines $x = a$ and $x = b$, then the area bounded by the curve, the axis, and the vertical lines is called the integral of f between a and b (draw a picture). The problem of finding tangents to curves is known as *differentiation* (finding a *derivative*), encoded in the relationship between infinitesimal changes along a curve, called *differentials,* another term due to Leibniz. These differentials, or *infinitesimals*, were a refinement of Cavalieri's indivisibles, and while their meaning and right to exist were vigorously attacked and debated, they were strikingly useful and successful, so concerns about their validity were subsumed by their efficacy. The essence of the Fundamental Theorem of Calculus of Leibniz and Newton is that integration and differentiation are inverse operations, and thus performing an integration is an antidifferentiation (i.e., inverse differentiation, or inverse tangent) problem.

There are two accompanying threads to follow as we explore whether the new calculus represented the complete solution to the area problem we set as our theme. First, how successful was it at solving problems, spawning new ones, and furthering mathematics and the physical and natural sciences? Here the answer is that the results were absolutely spectacular, quickly creating the "age of analysis" as the dominant force in mathematics and its applications through the entire eighteenth century. Second, what about the serious unresolved foundational questions? These difficult issues were in fact not resolved for more than another century, until the successes of the new analysis finally reached a point where the nagging foundational problems required resolution before further progress could be made.

Shortly after Leibniz's and Newton's discoveries, many other mathematicians embraced their methods, among them the distinguished Swiss Bernoulli brothers, Jakob (1654–1705) and Johann (1667–1748), and Leonhard Euler (1707–1783), Swiss mathematician and scientist extraordinaire, the most prolific author in all mathematical history. The Bernoullis translated geometrical and mechanical data into relationships between differentials, creating what we today call differential equations. Their results included lengths of curves of many new types, studying light caustics (curves occurring when light rays are reflected or refracted on curved surfaces, e.g., the beautiful curves formed on the surface of coffee with cream in a ceramic mug), and the form of sails blown by the wind. Two famous curve problems they solved were to find the shape of the catenary and the brachistochrone (the curve of shortest descent time for a bead moving down and across from one point to another, sliding without friction under the force of gravity, i.e., in some people's opinion the shape of the perfect ski slope). To their astonishment, the shape of the brachistochrone curve was the same as the solution to two other curve problems, the tautochrone and the cycloid. (See Simmons [157] for an exposition and wonderful exercises on these curves.)

Through the eighteenth century, calculus was enlarged into "mathematical analysis," with applications throughout the physical and natural sciences, primarily due to the vast and brilliant work of Euler. This involved a major shift away from working just with variable geometric quantities, towards working with functions to express specific relationships between variables. To Euler, "a function of a variable quantity is an analytic expression composed in whatever way of that variable and of numbers and constant quantities" [77]. In particular, he worked freely with infinite algebraic expressions, for instance beautiful and highly useful *infinite series* expressions like $\sin x = x - \frac{x^3}{3!} + \frac{x^5}{5!} - \frac{x^7}{7!} + \cdots$, which displays $\sin x$ as a polynomial of infinite degree, known today as a *power series* (see the Appendix (Section 3.8) for a brief introduction to infinite series). To Euler we owe our modern understanding of the power series representations of trigonometric, exponential, and logarithmic functions. His wizardry with infinite series makes inspiring reading [55]. Euler and others made incredible progress attacking an impressive variety of physical problems, for instance the physics and mathematics of elastic beams, vibrating strings, pendulum motion, projectile motion, water flow in pipes, and the motion of the moon, used for determining positions at sea.

During this exploration in what seemed like paradise, the creators of the new mathematical analysis often had to play fast and loose with the still rather shaky foundations of the calculus, but their intuition nonetheless guided them to wonderful results. To describe this vast expansion and its enormous impact on science could fill an entire book, but the reader may wish to look at [19, 81].

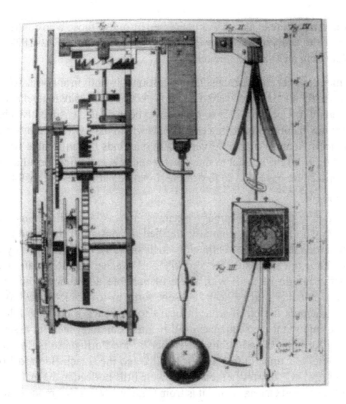

PHOTO 3.2. Huygens's cycloidal pendulum clock.

It slowly became clear that Euler's view of a function, as something given by an algebraic formula, was too narrow, especially for functions exhibiting periodic motion, such as for a vibrating string. An approach was developed that represented functions as infinite series built from various frequencies, by using the periodic trigonometric functions sine and cosine of multiples of the variable x, in place of the powers of x used for power series [75, 77]. These are named after one of their early nineteenth-century pioneers, Joseph Fourier (1768–1830), emerging particularly from his study of heat diffusion. Today Fourier series and their subsequent generalizations have become crucial tools in many diverse applications of mathematics to the physical world. But while the eighteenth-century explorations led to new types of functions and new ways to represent them, it also led directly back to the foundational questions left unresolved more than a century earlier, with the persistent appearance of vanishingly small magnitudes or differences.

The missing understanding underlying various coalescing challenges was "What is a limit?" i.e., what did it mean for a quantity that depends on a variable to approach a limiting value as the variable itself approaches a certain number. No conclusive answer had been proposed yet, as we will discuss in the section on Leibniz's work.

This frustrating state of affairs was expressed by one of those who contributed to bringing rigor to analysis, the Norwegian Niels Henrik Abel (1802–1829), when he complained in an 1826 letter about

> the tremendous obscurity which one unquestionably finds in analysis. It lacks so completely all plan and system that it is peculiar that so many men could have studied it. The worst of it is, it has never been treated stringently. There are very few theorems in advanced analysis which have been demonstrated in a logically tenable manner. Everywhere one finds this miserable way of concluding from the special to the general and it is extremely peculiar that such a procedure has led to so few of the so-called paradoxes [97, p. 947][1, vol. 2, pp. 263–65].

Despite a few eighteenth-century attempts, it was Augustin-Louis Cauchy (1789–1857) who finally substantially clarified the notion of limit and resolved many foundational difficulties, allowing analysis to develop further, by effectively obviating the necessity for slippery and evasive infinitesimals. Cauchy combined the notion of limit with those of variable and function to create a structure for calculus much like its present form in textbooks today. We now know that Cauchy's notion of what it means to approach a limiting value, and some of the important consequences of this understanding, were in essence independently formulated and published earlier by the Portuguese mathematician José Anastácio da Cunha (1744–1787) in 1782 and the Czech Bernard Bolzano (1781–1848) in 1817. However, their works were little noticed in the mathematical centers of France and Germany, and it is from Cauchy's work that modern analysis developed [93].

Cauchy presented his theory in lecture notes for his courses at the Ecole Polytechnique, in Paris, the new elite French institute of higher education and engineering. We will read selections leading to his formulation of the Fundamental Theorem of Calculus, illustrating shifts in point of view and more rigorous and unified understanding. For instance, we will see his thoroughly independent definitions of both the derivative and the integral of functions, unlike their circular intertwining in the work of Leibniz and Newton. Nonetheless, Cauchy did not break completely with the past, since he still attempted to interpret the infinitesimal within his new framework.

Cauchy's work synthesized the concepts and results of calculus with the deductive methods of ancient geometry, ushering in modern analysis, perhaps the largest and most fully developed branch of mathematics today.

Some of the most important unresolved questions of rigor still remaining in Cauchy's texts revolved around the notion of continuity. Intuitively, a function is continuous if its graph has no jumps; but does this mean, for instance, that all intermediate values are assumed? That is, if $f(x)$ is continuous for $a \leq x \leq b$, and if M lies between $f(a)$ and $f(b)$, is there necessarily some c between a and b for which $f(c) = M$? (Draw a picture!) This crucial property of continuity, upon which Cauchy's work relied, is called the Intermediate Value Theorem; its proof, provided independently by both Cauchy and Bolzano (see the set theory chapter), was still inadequate and demanded a deeper understanding of the real numbers themselves.

This emerged in the next fifty years, with the first actual definitions of the real numbers, founded only on the natural numbers, given by the German mathematicians Karl Weierstrass (1815–1897), Richard Dedekind (1831–1916), and Georg Cantor (1845–1918), who finally banished the seemingly useful fiction of the infinitesimal, in the so-called arithmetization of analysis. The proper foundation for the real numbers allowed a completely rigorous proof of the Intermediate Value Theorem, via the "completeness" property the real numbers possess, which ensures that they have no holes or gaps: specifically, any shrinking sequence of intervals must contain a point (real number) in common; i.e., one cannot shrink down to find an empty spot with no number there. One then also avoids the circularity inherent in statements like Cauchy's "an irrational number is the limit of the various fractions which provide values that approximate it more and more closely," in which his claim begs the very existence of irrational numbers.

This understanding of the real numbers progressed hand in hand with a broadening of the setting for analysis by mathematicians such as Bernhard Riemann (1826–1866) and Peter Lejeune-Dirichlet (1805–1859), for example, seeking detailed understanding of how to represent quite general functions by infinite series (see Appendix), and integrating functions with many, even infinitely many, points of discontinuity in an interval. Riemann slightly generalized Cauchy's definition of integration, in such a way that he could characterize precisely which functions could be integrated, while Dirichlet studied when a function could be represented by a Fourier series of trigonometric functions mentioned earlier. Thus began an intricate intertwining of the mutual development of the theories of integration, representations of functions by infinite series, and even of the very sense of what a function is. These efforts slowly grew beyond the original stimulus provided by the intuitive notion of area, but melded the emerging modern notions of continuity, discontinuity, variability, measurement of size of arbitrary sets of real numbers, integration, and functions and their representations by infinite series, ultimately becoming the modern synthesis we call real analysis. These nineteenth-century challenges in the development of real analysis directly motivated Georg Cantor's work on infinite sets featured in the set theory chapter.

By the turn of the twentieth century, researchers had discovered and tackled ever stranger functions. While some people wished simply to reject each new "pathology" discovered, others worked to incorporate them into the theory, and every time it emerged richer, more general, often more beautiful. The primary development, which set the stage for much of twentieth-century real analysis, was Henri Lebesgue's (1875–1941) introduction in 1902 of a totally new theory of integration, which we will explain and contrast with Cauchy's description of integration at the end of the section on Cauchy's work.

Meanwhile, other branches of analysis also emerged during the nineteenth century. Cauchy and his successors developed a beautiful theory of analysis for functions involving complex variables, i.e., those with both real and imaginary parts, with many applications today, e.g., to fluid flow and how an airplane wing provides lift. Other branches developed providing multivariable versions of all the important ideas; and today we even have a branch called functional analysis, in which functions themselves become mere points for other functions to be defined on, a metalevel analysis with applications in quantum physics.

At the close of the twentieth century, one of the hottest new fields in analysis is "wavelet theory," emerging from such applications as edge detection or texture analysis in computer vision, data compression in signal analysis or image processing, turbulence, layering of underground earth sediments, and computer-aided design. Wavelets are an extension of Fourier's idea of representing functions by superimposing waves given by sines or cosines. Since many oscillatory phenomena evolve in an unpredictable way over short intervals of time or space, the phenomenon is often better represented by superimposing waves of only short duration, christened wavelets. This tight interplay between current applications and a new field of mathematics is evolving so quickly that it is hard to see where it will lead even in the very near future [92].

We will conclude this chapter with an extraordinary modern twist to our long story. Recall that the infinitesimals of Leibniz, which had never been properly defined and were denigrated as fictional, had finally been banished from analysis by the successors of Cauchy in the nineteenth century, using a rigorous foundation for the real numbers. How surprising, then, that in 1960 deep methods of modern mathematical logic revived infinitesimals and gave them a new stature and role. In our final section we will read a few passages from the book *Non-Standard Analysis* [140] by Abraham Robinson (1918–1974), who discovered how to place infinitesimals on a firm foundation, and we will consider the possible consequences of his discovery for the future as well as for our evaluation of the past.

Exercise 3.1: Prove Hippocrates' theorem on the squaring of his lune.

Exercise 3.2: Research the history and eventual resolution of one of the three "classical problems" of antiquity.

Exercise 3.3: Find out what the "quadratrix of Hippias" is and how it was used in attempts to solve the problems of squaring the circle and trisecting the angle.

Exercise 3.4: Study Kepler's derivation [20, pp. 356 f.] of the area inside a circle. Critique his use of indivisibles. What are its strengths and weaknesses? Also study his matching of indivisibles to obtain the area inside an ellipse [20, pp. 356 f.]. Do you consider his argument valid? Why?

3.2 Archimedes' Quadrature of the Parabola

Archimedes (c. 287–212 B.C.E.) was the greatest mathematician of antiquity, and one of the top handful of all time. His achievements seem astounding even today. The son of an astronomer, he spent most of his life in Syracuse, on the island of Sicily, in present-day southern Italy, except for a likely period in Alexandria studying with successors of Euclid. In addition to spectacular mathematical

PHOTO 3.3. Archimedes.

achievements, his reputation during his lifetime derived from an impressive array of mechanical inventions, from the water snail (a screw for raising irrigation water) to compound pulleys, and fearful war instruments described in the Introduction. Referring to his principle of the lever, Archimedes boasted, "Give me a place to stand on, and I will move the earth." When King Hieron of Syracuse heard of this and asked Archimedes to demonstrate his principle, he demonstrated the efficacy of his pulley systems based on this law by easily pulling single-handedly a three-masted schooner laden with passengers and freight [93]. One of his most famous, but possibly apocryphal, exploits was to determine for the king whether a gold-smith had fraudulently alloyed a supposedly gold crown with cheaper metal. He is purported to have realized, while in a public bath, the principle that his floating body displaced exactly its weight in water, and, realizing that he could use this to solve the problem, rushed home naked through the streets shouting "Eureka, Eureka" (I have found it!).

The treatises of Archimedes contain a wide array of area, volume, and center of gravity determinations, including virtually all the best-known formulas taught in high school today. As mentioned in the Introduction, he was so pleased with his results about the sphere that he had one of them inscribed on his gravestone: The volume of a sphere is two-thirds that of the circumscribed cylinder, and astonishingly, the same ratio holds true for their surface areas. It was typical at the time to compare two different geometric objects in this fashion, rather than using formulas as we do today. Archimedes also laid the mathematical foundation for the fields of statics and hydrodynamics and their interplay with geometry, and frequently used intricate balancing arguments. A fascinating treatise of his on a different topic is *The Sandreckoner*, in which he numbered the grains of sand needed to fill the universe (i.e., a sphere with radius the estimated distance to the sun), by developing an effective system for dealing with large numbers. Even though he calculated in the end that only 10^{63} grains would be needed, his system could actually calculate with numbers as enormous as $\left(\left(10^8\right)^{10^8}\right)^{10^8}$. Archimedes even modeled the universe with a mechanical planetarium incorporating the motions of the sun, the moon, and the "five stars which are called the wanderers" (i.e., the known planets) [42].

We will examine two remarkably different texts Archimedes wrote on finding the area of a "segment" of a parabola. A *segment* is the region bounded by a parabola and an arbitrary line cutting across the parabola. The portion of the cutting line between the two intersection points is called a chord, and forms what Archimedes calls the base of the segment. He states his beautiful result in a letter to Dositheus, a successor of Euclid's in Alexandria, prefacing his treatise *Quadrature of the Parabola* [3, pp. 233–34].

ARCHIMEDES to Dositheus greeting.

When I heard that Conon,[2] who was my friend in his lifetime, was dead, but that you were acquainted with Conon and withal versed in geometry, while I grieved for the loss not only of a friend but of an admirable mathematician, I set myself the task of communicating to you, as I had intended to send to Conon, a certain geometrical theorem which had not been investigated before but has now been investigated by me, and which I first discovered by means of mechanics and then exhibited by means of geometry. Now some of the earlier geometers tried to prove it possible to find a rectilineal area equal to a given circle and a given segment of a circle.... But I am not aware that any one of my predecessors has attempted to square the segment bounded by a straight line and a section of a right-angled cone [a parabola], of which problem I have now discovered the solution. For it is here shown that every segment bounded by a straight line and a section of a right-angled cone [a parabola] is four-thirds of the triangle which has the same base and equal height with the segment, and for the demonstration of this property the following lemma is assumed: that the excess by which the greater of (two) unequal areas exceeds the less can, by being added to itself, be made to exceed any given finite area. The earlier geometers have also used this lemma; for it is by the use of this same lemma that they have shown that circles are to one another in the duplicate ratio of their diameters, and that spheres are to one another in the triplicate ratio of their diameters, and further that every pyramid is one third part of the prism which has the same base with the pyramid and equal height; also, that every cone is one

---

[2] Another successor of Euclid's.

third part of the cylinder having the same base as the cone and equal height they proved by assuming a certain lemma similar to that aforesaid. And, in the result, each of the aforesaid theorems has been accepted no less than those proved without the lemma. As therefore my work now published has satisfied the same test as the propositions referred to, I have written out the proof and send it to you, first as investigated by means of mechanics, and afterwards too as demonstrated by geometry. Prefixed are, also, the elementary propositions in conics which are of service in the proof. Farewell.

Archimedes provides several points of view in demonstrating his result. In *Quadrature of the Parabola* he gives two proofs representing formal Greek methods, while yielding little insight into how the result might have been discovered. In the second, which he calls "geometrical," we will see the method of exhaustion in action. While Archimedes develops and uses many beautiful and fascinating features of parabolas in order to prove his result, most of which are unfamiliar to us today, we will focus primarily on how he combines these with the method of exhaustion, encouraging the reader to explore the geometric underpinnings.

Our other text, *The Method*, will reveal how Archimedes actually discovered his result by an imaginary physical balancing technique using indivisibles. While he considered this only heuristic, and not acceptable as proof, we shall see that it foreshadows later methods of the calculus by about two thousand years.

A parabola is an important curve in part because it can be described by a number of equivalent but very different properties, each simple and aesthetically pleasing. This reflects the fact that parabolas arise in many mathematical and physical situations. While the reader is probably familiar with a parabola in some form, it is surprising how different our typical view of it is today from that of two thousand years ago.

Parabolas are one of three types of curves (along with hyperbolas and ellipses) first studied as certain cross-sections created by a plane slicing through a cone, hence the name "conic sections" for these curves. While their discovery is attributed to the geometer Menaechmus around 350 B.C.E., we do not know how the relationship between their purely planar properties and their description as conic sections was discovered [43, pp. 56–57].

Greek mathematicians derived a planar "symptom" for each parabola [43, pp. 57 f.], a characteristic relation between the coordinates of any point on the curve, using measurements along a pair of coordinate directions (not necessarily mutually perpendicular!) to describe the positions of points. In modern algebraic symbolism, the symptom becomes an equation for the curve. It is fascinating to study how this was probably done for the various conic sections [43, pp. 57 f.] (Exercise 3.5). Of course, we know that in an appropriate modern (perpendicular) Cartesian coordinate system, the equation is simply $py = x^2$ (p a constant depending on how far from the vertex we slice the cone). One of the astonishing things Archimedes knew is that many features of a parabola, including its symptom, hold for oblique axes as well (Exercises 3.6, 3.7) [43, pp. 57 f.]. This offers a hint at the more modern subject of affine geometry [34].

We are ready to read selections from Archimedes' treatise *Quadrature of the Parabola* [3, pp. 233–37, 246–52]. The approach of the proof is to inscribe polygons inside the parabolic segment to approximate its area, and then use the method of exhaustion to confirm an exact, not merely approximate, value for its area. By a "tangent" to a parabola, Archimedes means a line touching it in exactly one point. He assumes that each point on a parabola has exactly one tangent line containing it (Exercise 3.8). He uses the word "diameter" to refer to any line parallel to the axis of symmetry of the parabola, and "ordinate" to refer to coordinate measurement along the oblique coordinate in the direction parallel to the tangent.

Archimedes, from

Quadrature of the Parabola

Definition. In segments bounded by a straight line and any curve I call the straight line the **base**, and the **height** the greatest perpendicular drawn from the curve to the base of the segment, and the **vertex** the point from which the greatest perpendicular is drawn.

Proposition 20.

If Qq be the base, and P the vertex, of a parabolic segment, then the triangle PQq is greater than half the segment PQq. [See Figure 3.2]

For the chord Qq is parallel to the tangent[3] at P, and the triangle PQq is half the parallelogram formed by Qq, the tangent at P, and the diameters through Q, q.

Therefore the triangle PQq is greater than half the segment.

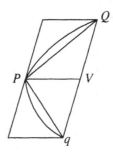

FIGURE 3.2. Proposition 20.

---

[3] Read Propositions 1 and 18 of Archimedes' treatise [3], and follow Exercise 3.9, to see why this beautiful fact holds.

Cor. It follows that *it is possible to inscribe in the segment a polygon such that the segments left over are together less than any assigned area.*

This corollary refers to one of the cornerstones of the method of exhaustion; we will postpone discussing it until Archimedes elaborates in Proposition 24.

Proposition 22.

If there be a series of areas A, B, C, D, ... each of which is four times the next in order, and if the largest, A, be equal to the triangle PQq inscribed in a parabolic segment PQq and having the same base with it and equal height, then

$$(A + B + C + D + \cdots) < (\text{area of segment } PQq).$$

For, since $\triangle PQq = 8\triangle PRQ = 8\triangle Pqr$ (see Figure 3.3), where R, r are the vertices of the segments cut off by PQ, Pq, then as in the last proposition,[4]

$$\triangle PQq = 4(\triangle PQR + \triangle Pqr).$$

Therefore, since $\triangle PQq = A$,

$$\triangle PQR + \triangle Pqr = B.$$

In like manner we prove that the triangles similarly inscribed in the remaining segments are together equal to the area C, and so on.

Therefore $A + B + C + D + \cdots$ is equal to the area of a certain inscribed polygon, and is therefore less than the area of the segment.

Proposition 23.

Given a series of areas A, B, C, D, ..., Z, of which A is the greatest, and each is equal to four times the next in order, then

$$A + B + C + \cdots + Z + \tfrac{1}{3}Z = \tfrac{4}{3}A.$$

We ask the reader to prove this algebraic result (Exercise 3.11).

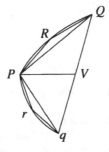

FIGURE 3.3. Proposition 22.

---

[4]Study Propositions 3, 19, and 21.

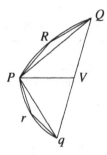

FIGURE 3.4. Proposition 24.

The finale now combines Propositions 20, 22, and 23 to prove the theorem.

Proposition 24.

Every segment bounded by a parabola and a chord Qq is equal to four-thirds of the triangle which has the same base as the segment and equal height.
Suppose

$$K = \tfrac{4}{3} \Delta PQq,$$

where P is the vertex of the segment; and we have then to prove that the area of the segment is equal to K. [See Figure 3.4]
For, if the segment be not equal to K, it must either be greater or less.

I. Suppose the area of the segment greater than K.
If then we inscribe in the segments cut off by PQ, Pq triangles which have the same base and equal height, i.e., triangles with the same vertices R, r as those of the segments, and if in the remaining segments we inscribe triangles in the same manner, and so on, we shall finally have segments remaining whose sum is less than the area by which the segment PQq exceeds K.
Therefore the polygon so formed must be greater than the area K; which is impossible, since [Prop. 23]

$$A + B + C + \cdots + Z < \tfrac{4}{3}A,$$

where

$$A = \Delta PQq.$$

Thus the area of the segment cannot be greater than K.

II. Suppose, if possible, that the area of the segment is less than K.
If then $\Delta PQq = A$, $B = \tfrac{1}{4}A$, $C = \tfrac{1}{4}B$, and so on, until we arrive at an area X such that X is less than the difference between K and the segment, we have

$$A + B + C + \cdots + X + \tfrac{1}{3}X = \tfrac{4}{3}A \qquad \text{[Prop. 23]}$$
$$= K.$$

Now, since K exceeds $A + B + C + \cdots + X$ by an area less than X, and the area of the segment by an area greater than X, it follows that

$$A + B + C + \cdots + X > \text{(the segment)};$$

which is impossible, by Prop. 22 above.

Hence the segment is not less than K.

Thus, since the segment is neither greater nor less than K,

$$\text{(area of segment } PQq) = K = \tfrac{4}{3}\Delta PQq.$$

Early in the proof, Archimedes asserts that if the (area of the) segment does not equal K, it must be either greater or less, and he proceeds to rule out both these possibilities. This was a very common method of proof in his time, known now as double *reductio ad absurdum*. (What does this Latin phrase mean, and how does it describe the method?)

In parts I and II, polygons of computable area are inscribed that are sufficiently close in area to the segment to provide the needed contradictions to the possibility that the area of the segment could be greater or less than K, even though sufficiently close is necessarily of an arbitrary, unknown nature. This is the essence of the method of exhaustion; the area is being exhausted by the polygons. This is really an inappropriate term, though, since any polygon necessarily leaves some area unfilled; the method was designed precisely to get around this problem without confronting the completing of the filling process. Comprehending the connection between the infinite filling process and the area of the entire segment became the goal of a two thousand year struggle, only resolved in the nineteenth century.

How does Archimedes know that his process of including ever more triangles will ultimately provide a polygon differing in area from the segment by as little as needed? This is the Corollary he noted after Proposition 20. How does he know that Proposition 20 means that the area left over can be made as small as needed, in his words "less than any assigned area"? We find the answer by rereading Archimedes' introductory letter!

He tells Dositheus that "the following lemma is assumed" and also "a certain lemma similar to the aforesaid"; and that these lemmas were used by "earlier geometers" to prove their results on areas of circles and volumes of spheres and cones. The second lemma he refers to here had been stated by Euclid in *Elements* [51, Vol. 3, p. 14], as Proposition 1 in Book X:

> Two unequal magnitudes being set out, if from the greater there be subtracted a magnitude greater than its half, and from that which is left a magnitude greater than its half, and if this process be repeated continually, there will be left some magnitude which will be less than the lesser magnitude set out.

Clearly, this is exactly what Archimedes uses to obtain his Corollary to Proposition 20 and thus his claim in Proposition 24. Intuitively, it says that by successively halving a magnitude, we can eventually, but always after some finite number of halvings, make it smaller than some previously assigned magnitude. The previously assigned magnitude, although perhaps small and unknown, is fixed, and this is what makes it possible to eventually undershoot it, since it does not recede as the halving process progresses.

This assumed lemma on the part of Archimedes is called the "dichotomy principle." While it may seem quite reasonable, this is deceptive, since its great power actually amounts to ruling out the existence of infinitesimally small magnitudes, i.e., those that are smaller than any fraction of a unit and yet still not zero. The dichotomy principle was an assumption about the fundamental nature of magnitudes, which Archimedes could not prove; it was only set on a firm foundation much later, when a truly satisfactory definition for real numbers was discovered (Exercise 3.12).

To summarize, Archimedes combined the exhaustion method with a deep geometric understanding of parabolas and a clever summation of a series of terms with identical successive ratios (today called a "geometric" series; see the chapter Appendix), to demonstrate that the exact area of a parabolic segment is four-thirds of a certain triangle (Exercises 3.13, 3.14). His presentation is an example of "synthesis," in which simple pieces are systematically put together to yield something more complex. It leaves us puzzled, though, about how Archimedes discovered his result. Our second text will actually reveal the answer to this riddle.

Exercise 3.5: Derive the symptom (equation) of the parabola using modern algebraic notation, and compare with the spirit of how the ancients may have done it [43, pp. 57 f.].

Exercise 3.6: Learn about the Greek method of "application of areas," and explain how Apollonius used it in enlarging the notion of conic sections [20, 127, 173][43, pp. 51 f., 61 f.]. Explain how this accounts for the terms parabola, hyperbola, ellipse.

Exercise 3.7: Learn about how Apollonius expanded the idea of conic sections to cutting a right-angled cone at any angle and to oblique cones, and still (surprisingly) obtained the same curves as before [43, pp. 59 f.]. Describe the details in your own words using modern notation.

Exercise 3.8: Show that each point on a parabola has exactly one tangent line. Use just the symptom. Any use of calculus (e.g., a derivative) would of course be cheating by a couple of millennia. Hint: Archimedes' Proposition 2 from *Quadrature of the Parabola* [3] should suggest which line to focus your attention on.

Exercise 3.9: Prove Propositions 1, 2, and 3 of Archimedes' treatise *Quadrature of the Parabola* [3]. We will outline one path to doing so, from a modern point of view, although you may wish to find your own.

- From Exercise 3.8 we have a complete description of the tangent line to a parabola at any point.
- From the equation $py = x^2$ for the symptom of a parabola, verify Proposition 3 directly algebraically. Go one step further to obtain the equation for the symptom of the parabola using the new oblique coordinates. Notice how the constant p changes in the new coordinates, and how this change depends on the

angle of the relevant tangent line for the new coordinates. You might wish to do this part by formally introducing the new coordinates, seeing how to convert between old and new (this particular type of change is called an "affine" change of coordinates), and then transforming the equation for the symptom into the new coordinates.

- Now either prove the general oblique form of Propositions 1 and 2 by using Proposition 3 and some algebra, or carefully create an argument that because the symptom is true for both oblique and right-angled coordinates from Proposition 3, the claims of Propositions 1 and 2 will follow for oblique axes, since we already know they are true for the standard axes.

Exercise 3.10: Consider the situation of Proposition 2 from *Quadrature of the Parabola* [3], but using only standard right-angled coordinate axes. The length of *TV* is called the "subtangent" because it is the length of the portion of the axis corresponding to that portion of the tangent line from the point of tangency to the axis. We have seen that the length of the subtangent is twice the coordinate *PV*. Another important line is the "normal," the line at a right angle to the tangent line at *Q*. Then the "subnormal" is defined by analogy to the subtangent. Find the subnormal. What does the subnormal surprisingly *not* depend on?

Exercise 3.11:

- Derive your own algebraic proof of Proposition 23. Generalize this to any series with arbitrary ratio between the terms, rather than always four to one.
- Archimedes seems to have proven Proposition 23 only for a sum of twenty-six areas $A, \ldots, Z$. Discuss. How could we phrase this more generally today? Why didn't he do so?
- Can you obtain from Proposition 23 the value of a certain infinite sum? Archimedes' method of proof is designed to avoid actually doing this, since infinite sums were considered unrigorous.

Exercise 3.12: The dichotomy principle has an equivalent form as the first lemma in Archimedes' letter, asserting nonexistence of infinitely large magnitudes, which today is called the postulate of Eudoxus. Study it and its use [84, 85], and show that the dichotomy principle and the postulate of Eudoxus are equivalent (Hint: Look at Euclid, Book X, Proposition 1 for help.)

Exercise 3.13: Compare and contrast Archimedes' proof of his result on the area of a circle [3, pp. 91 f.] with the exhaustion proof he has given for the area of a segment of a parabola. Note where he uses yet another "assumption" in that proof, namely Assumption 6 in *On the Sphere and Cylinder, I* [3, p. 4]. Discuss why he must make this assumption, and what would be needed to elevate this assumption to something one could actually prove.

Exercise 3.14: Archimedes proves that the area of a segment of a parabola is four-thirds that of a certain triangle. But this is of limited usefulness unless one can determine the dimensions of the triangle for a given segment, in order to compute its area. Show how this can always be done.

3.3   Archimedes' Method

Our second source comes from *The Method of Treating Mechanical Problems*, an extraordinary manuscript with an astonishing history. While it was alluded to in ancient commentaries, it was thought nonexistent or irretrievably lost until its remarkable rediscovery by the Danish scholar J. L. Heiberg at the turn of the twentieth century [43, pp. 44 f.], [3, Supplement].

Heiberg's attention was drawn to an 1899 report about a palimpsest with originally mathematical contents in the library of the monastery of the Holy Sepulchre, in Jerusalem. A palimpsest is a parchment, tablet, or other surface, that has been written on more than once, with the previous text(s) imperfectly erased, and therefore still visible. Of course, in times past, before paper was plentiful, this was a common practice. A few lines of erased text quoted in the report convinced Heiberg that the underlying text was by Archimedes, and he was able to examine and photograph the parchment in Constantinople. It contained an Archimedean text written in a tenth century hand, and an imperfect attempt had been made to wash it off and replace it with a religious text in the twelfth to fourteenth century. Heiberg succeeded in deciphering most of the underlying manuscript, which contains versions of previously known works by Archimedes, and the almost complete text of the long lost *Method*. It is interesting that this only extant copy was first preserved by religious efforts, then obliterated for another religious purpose, and now through an incredible stroke of luck is available to us today.[5]

The significance of the rediscovered manuscript is explained by two distinguished scholars.

E.J. Dijksterhuis writes:

Greek mathematics is characterized—and in this respect, too, it founded a tradition which was to last down to our own time—by a care of the form of the mathematical argument which, superficially viewed, seems almost exaggerated. It demands the inexorably proceeding, irrefutably persuading sequence of logical conclusions constituting the synthetic method of demonstration, but to this it sacrifices the reader's wish to gain also an insight into the method by which the result was first discovered. It is this wish, however, which Archimedes meets in his *Method:* he will reveal how he himself, long before he knew how to prove his theorems, became convinced of their truth [43, p. 315].

And T.L. Heath says:

[H]ere we have a sort of lifting of the veil, a glimpse of the interior of Archimedes' workshop as it were. He tells us how he discovered certain theorems in quadrature and cubature, and he is at the same time careful to insist on the difference between (1) the means which may be sufficient to suggest the

---

[5] As this book goes to print in October, 1998, the palimpsest has reappeared after almost a century in the hands of private collectors, and is being auctioned at Christie's in New York for about one million dollars.

truth of theorems, although not furnishing scientific proofs of them, and (2) the rigorous demonstrations of them by irrefragable geometrical methods which must follow before they can be finally accepted as established; to use Archimedes' own terms, the former enable theorems to be *investigated* but not be *proved* [3, Supplement, pp. 6–7].

Archimedes' explanation of how he discovered the area of a parabolic segment involves two ideas very different from Eudoxean exhaustion: a "mechanical balancing" method, and "summation of indivisibles." Let us begin with Archimedes' prefatory letter:

Archimedes to Eratosthenes[6] greeting.

I sent you on a former occasion some of the theorems discovered by me, merely writing out the enunciations and inviting you to discover the proofs, which at the moment I did not give. The enunciations of the theorems which I sent were as follows....

Seeing moreover in you, as I say, an earnest student, a man of considerable eminence in philosophy, and an admirer [of mathematical inquiry], I thought fit to write out for you and explain in detail in the same book the peculiarity of a certain method, by which it will be possible for you to get a start to enable you to investigate some of the problems in mathematics by means of mechanics. This procedure is, I am persuaded, no less useful even for the proof of the theorems themselves; for certain things first became clear to me by a mechanical method, although they had to be demonstrated by geometry afterwards because their investigation by the said method did not furnish an actual demonstration. But it is of course easier, when we have previously acquired, by the method, some knowledge of the questions, to supply the proof than it is to find it without any previous knowledge. This is a reason why, in the case of the theorems the proof of which Eudoxus was the first to discover, namely that the cone is a third part of the cylinder, and the pyramid of the prism, having the same base and equal height, we should give no small share of the credit to Democritus, who was the first to make the assertion with regard to the said figure though he did not prove it. I am myself in the position of having first made the discovery of the theorem now to be published [by the method indicated], and I deem it necessary to expound the method partly because I have already spoken of it and I do not want to be thought to have uttered vain words, but equally because I am persuaded that it will be of no little service to mathematics; for I apprehend that some, either of my contemporaries or of my successors, will, by means of the method when once established, be able to discover other theorems in addition, which have not yet occurred to me.

---

[6]Eratosthenes was a great scholar in many fields, including mathematics and astronomy. He made the most accurate calculation in antiquity of the circumference of the earth, based on measurements of the sun's shadow at different latitudes. After many years in Athens, Eratosthenes was appointed chief librarian at the Museum/University in Alexandria, and it was to him that Archimedes sent *The Method*.

First then I will set out the very first theorem which became known to me by means of mechanics, namely that

Any segment of a section of a right-angled cone (i.e., a parabola) is four-thirds of the triangle which has the same base and equal height, and after this I will give each of the other theorems investigated by the same method. Then, at the end of the book, I will give the geometrical [proofs of the propositions].

Archimedes proceeds to Proposition 1 of *The Method*, his parabolic area result.

The Method makes heavy use of balancing principles from mechanics (the movement of bodies under the action of forces). Here Archimedes leaves pure mathematics and appeals to natural science considered mathematically, claiming to draw mathematical conclusions from the physical science of weights, centers of gravity, and equilibria. He was the first to establish a systematic connection between mathematics and mechanics, developing much of that portion of mechanics today called statics. From intuitively acceptable postulates, he developed a variety of results on static equilibria, most importantly the

Principle of the Lever

Two magnitudes balance at distances reciprocally proportional to the magnitudes [3, *On the Equilibrium of Planes*, Book I, Propositions 6,7].

To apply this principle we must understand how to interpret the appropriate point of positioning for an arbitrary geometric magnitude, called its "center of gravity" by Archimedes; intuitively, it is the point upon which the magnitude itself will balance in equilibrium. The foundational issues involved in properly interpreting this idea are intricate and fascinating (Exercise 3.15). It is not hard to believe from symmetry that the center of gravity of a line segment is the midpoint of the segment, or that the center of gravity of a rectangle is the point where the diagonals intersect. But what about a triangle? This is the type of object Archimedes studies, and he proves (Book I, Proposition 13) that its center of gravity is the point where the median lines intersect (recall that a median connects a vertex to the midpoint of the opposite side). Archimedes mentions in Book I, Proposition 15, that this point is two-thirds of the way along each median (Exercise 3.16). Now we are ready for *The Method*.

Archimedes, from
The Method of Treating of Mechanical Problems

Proposition 1.

Let ABC be a segment of a parabola bounded by the straight line AC and the parabola ABC, and let D be the middle point of AC. [See Fig. 3.5] Draw the straight line DBE parallel to the axis of the parabola and join AB, BC.

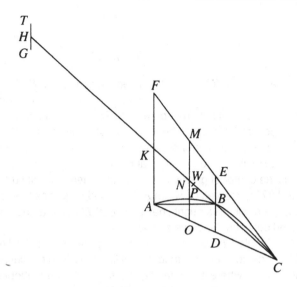

FIGURE 3.5. Proposition 1.

Then shall the segment ABC be $\frac{4}{3}$ of the triangle ABC.

From A draw AKF parallel to DE, and let the tangent to the parabola at C meet DBE in E and AKF in F. Produce CB to meet AF in K, and again produce CK to H, making KH equal to CK.

Consider CH as the bar of a balance, K being its middle point.

Let MO be any straight line parallel to ED, and let it meet CF, CK, AC in M, N, O and the curve in P.

Now, since CE is a tangent to the parabola and CD the semi-ordinate,

$$EB = BD;^7$$

"for this is proved in the Elements [of Conics]*."

Since FA, MO are parallel to ED, it follows that

$$FK = KA, MN = NO.$$

Now, by the property of the parabola, "proved in a lemma,"

$$
\begin{aligned}
MO : OP \quad &= CA : AO &\quad \text{[cf. \textit{Quadrature of Parabola}, Prop. 5]}^8 \\
&= CK : KN &\quad \text{[Eucl. VI. 2]} \\
&= HK : KN.
\end{aligned}
$$

Take a straight line TG equal to OP, and place it with its center of gravity at H, so that $TH = HG$; then, since N is the center of gravity of the straight line MO,

^{7}Read Propositions 1 and 2 in *Quadrature of the Parabola* to see why this is true.

*I.e., the works on conics by Aristaeus and Euclid....

^{8}This surprising result is explored in Exercise 3.17.

and

$$MO : TG = HK : KN,$$

it follows that TG at H and MO at N will be in equilibrium about K.

[*On the Equilibrium of Planes*, I. 6, 7]

Similarly, for all other straight lines parallel to DE and meeting the arc of the parabola, (1) the portion intercepted between FC, AC with its middle point in KC and (2) a length equal to the intercept between the curve and AC placed with its center of gravity at H will be in equilibrium about K.

Therefore K is the center of gravity of the whole system consisting (1) of all the straight lines as MO intercepted between FC, AC and placed as they actually are in the figure and (2) of all the straight lines placed at H equal to the straight lines as PO intercepted between the curve and AC.

And, since the triangle CFA is made up of all the parallel lines like MO, and the segment CBA is made up of all the straight lines like PO within the curve, it follows that the triangle, placed where it is in the figure, is in equilibrium about K with the segment CBA placed with its center of gravity at H.

Divide KC at W so that $CK = 3KW$;

then W is the center of gravity of the triangle ACF; "for this is proved in the books on equilibrium."

[Cf. *On the Equilibrium of Planes* I. 15]

Therefore $\triangle ACF$:(segment ABC) $=$ $HK : KW$
$=$ $3 : 1$.
Therefore segment ABC $=$ $\frac{1}{3}\triangle ACF$.
But $\triangle ACF$ $=$ $4\triangle ABC$.
Therefore segment ABC $=$ $\frac{4}{3}\triangle ABC$.

Now the fact here stated is not actually demonstrated by the argument used; but that argument has given a sort of indication that the conclusion is true. Seeing then that the theorem is not demonstrated, but at the same time suspecting that the conclusion is true, we shall have recourse to the geometrical demonstration which I myself discovered and have already published.

Archimedes' "mechanical" argument is beautiful in its simplicity, and displays the mark of a genius, expert at the principle of the lever and centers of gravity. He was clearly comfortable with the idea that a region is made up of all its parallel lines, and his *Method* depends precisely on the delicate interplay between the summation of indivisibles and the application of the Principle of the Lever to both the indivisibles and the regions.

Archimedes viewed his method only as a heuristic discovery tool, reverting to geometric Eudoxean exhaustion when he published a rigorous proof. What remains tantalizingly unclear even today is whether it is the mechanical (balancing) or indivisible aspect of the method that taints it as a method of rigorous proof by classical Greek standards. Modern scholars provide compelling arguments and evidence in several directions on these questions [43, pp. 319 ff.], [99]; this is highly recommended reading (Exercise 3.18).

With hindsight, the challenge of making *The Method* rigorous lies not so much in mathematicizing the apparent appeal to physics in the mechanical aspect of the treatise, for this is easily done. Rather, the use of indivisibles is the sticky point. Greek mathematics had struggled with this in the broader context of the infinite for several centuries, and it would be many more centuries before these questions would be tackled on a new level (Exercise 3.19).

Exercise 3.15: Investigate the controversy over the assumptions Archimedes makes in developing and using his "Principle of the Lever" [43, pp. 286 ff.].

Exercise 3.16: Prove that the medians of a triangle intersect two-thirds of the way along each median.

Exercise 3.17: Learn about Proposition 5 of *The Method* [3], and prove it directly from Propositions 1, 2, 3 of *Quadrature of the Parabola* [3]. Does modern coordinate terminology help?

Exercise 3.18: Investigate the controversy today over Archimedes' use of mechanical and indivisible methods.

Exercise 3.19: The Fundamental Theorem of Calculus, whose seventeenth-century proof by Leibniz we will study later, enables one to calculate the area "under" a portion of a parabola. Considering the parabola $y = kx^2$ and the region bounded by this curve, the x-axis, and the verticals $x = a$ and $x = b$, the Fundamental Theorem will yield the result $k(b^3 - a^3)/3$ for its area. Obtain this value directly from Archimedes' theorem *Quadrature of the Parabola*, without using Leibniz's Fundamental Theorem of Calculus.

3.4 Cavalieri Calculates Areas of Higher Parabolas

Bonaventura Cavalieri was one of several early-seventeenth-century mathematicians who developed their own methods of using indivisibles, combined with emerging algebraic knowledge, to calculate many new areas and volumes.

As a boy Cavalieri entered the Gesuati religious order (not to be confused with the Jesuits) [42], [157, p. 106], and through a monk who had studied with Galileo he was introduced to geometry, studied at the University of Pisa, and became a lifelong disciple of Galileo's. For a number of years he was prior of the monastery in Parma, and by 1627 he announced to Galileo that he had finished a book on indivisibles, *Geometria indivisibilibus continuorum nova quadam ratione promota* (Geometry by indivisibles of the continua advanced by a new method). Galileo successfully recommended him for a professorship at the University of Bologna, saying "few, if any, since Archimedes, have delved as far and as deep into the science of geometry" [42]. "He has discovered a new method for the study of mathematical truths" [157, p. 106].

Although all the writings of Archimedes known to mathematicians of that era were based on the method of exhaustion (recall that *The Method* lay undiscovered until the twentieth century), they believed that the ancients must have had another method of discovery. Cavalieri's contemporary Evangelista Torricelli wrote:

> I should not dare affirm that this geometry of indivisibles is actually a new discovery. I should rather believe that the ancient geometricians availed themselves of this method in order to discover the more difficult theorems, although in their demonstration they may have preferred another way, either to conceal the secret of their art or to afford no occasion for criticism by invidious detractors. Whatever it was, it is certain that this geometry represents a marvelous economy of labor in the demonstrations and establishes innumerable, almost inscrutable, theorems by means of brief, direct, and affirmative demonstrations, which the doctrine of the ancients was incapable of. The geometry of indivisibles was indeed, in the mathematical briar bush, the so-called royal road, and one that Cavalieri first opened and laid out for the public as a device of marvelous invention [42].

Cavalieri systematized the use of indivisibles, and considered it a valid way of obtaining new results, without need for independent demonstration by exhaustion. He never attempted to define exactly what his indivisibles were, or to explain exactly how an area or volume was made up of them. This spared him to some extent from becoming embroiled in philosophical arguments over the nature of the infinite [8, 21]. Rather, he developed general principles and methods for their use, such as

Cavalieri's Principle

> If two plane figures cut by a set of parallel straight lines intersect, on each of these straight lines, equal chords, the two figures are equivalent [i.e., of equal area]....

> Similarly, in space: if the sections of two solids obtained by means of planes that are parallel to each other are equivalent two by two, the two solids are equivalent [i.e., of equal volume]... [42].

The reader should draw some pictures illustrating his principle and explain why it is believable. In addition to the exercises we provide illustrating the use of Cavalieri's Principle, more may be found in [157].

On the other hand, Cavalieri was also well aware of the limitations of indivisibles. In a letter to Torricelli he observed (Figure 3.6) that if one takes a nonisosceles triangle ABC with altitude AD, and considers an arbitrary line PQ parallel to the base BC, this produces a pair of lines PR and QS to the base, parallel to AD. Thus there is a one-to-one correspondence between each of the indivisibles PR, which taken together make up triangle ABD, and each of the indivisibles QS of the same length, which together form ACD. These areas thus appear equal, which is of course absurd. (Does this argument make bona fide use of Cavalieri's Principle?)

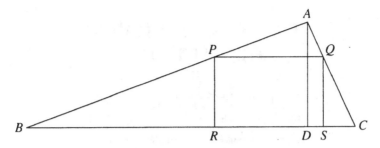

FIGURE 3.6. Cavalieri's triangle paradox.

Cavalieri explains away his paradox by considering the indivisibles *PR*, *QS* to be like the threads of a fabric, and points out that if *AC* contains, say, 100 points, and if *AB* contains, say, 200 points, then *ABD* has more threads than *ACD*. This appears to give his indivisibles actual width, making them two-dimensional (and thus not truly "indivisible"), but he did not manage to pursue this idea consistently [166, p. 218]. We will see that the later work of Leibniz begins the resolution of this mystery.

These unresolved issues provided fertile ground for others to attack Cavalieri's method, but his approach was destined to be refined and its validity essentially vindicated. Leibniz would later say, "In the sublimest of geometry, the initiators and promoters who performed yeomen's task in that field were Cavalieri and Torricelli" [42].

Cavalieri's work on indivisibles was refined in his second book, *Exercitationes Geometricae Sex* (Six Geometrical Exercises) [32], from which we now study his quadrature of a cubic curve [166, pp. 215–216]. Note the distinction in meaning between his words "together with" and "with," and also that "a.l., a.s., a.c." refer to "all lines, all squares, all cubes," with "all" meaning the sum total of the relevant indivisibles to form a plane figure, or solid, or something higher-dimensional in the case of all cubes. In considering "all cubes" Cavalieri was willing to go outside the constraints of three geometrical dimensions.

Bonaventura Cavalieri, from
Six Geometrical Exercises

Proposition 21. All cubes of the parallelogram AD [Figure 3.7] *are the quadruple of all cubes of either triangles ACF or FDC.*

All cubes of parallelogram *AD* are equal to a.c. of [the line *NH* of] triangle *ACF* together with a.c. of [the lines *HE* of] triangle *FDC* together with three times a.l. of triangle *ACF* with a.s. of [the lines of] triangle *FDC*, together with three times a.l. of triangle *FDC* with a.s. of [the lines of] triangle *ACF*.

Now a.c. of *AD* are the product of a.l. *AD* by a.s. *AD*, and this is to the product of a.l. of *AD* by a.s. of triangle *FDC* as a.s. of the parallelogram *AD* is to a.s. of

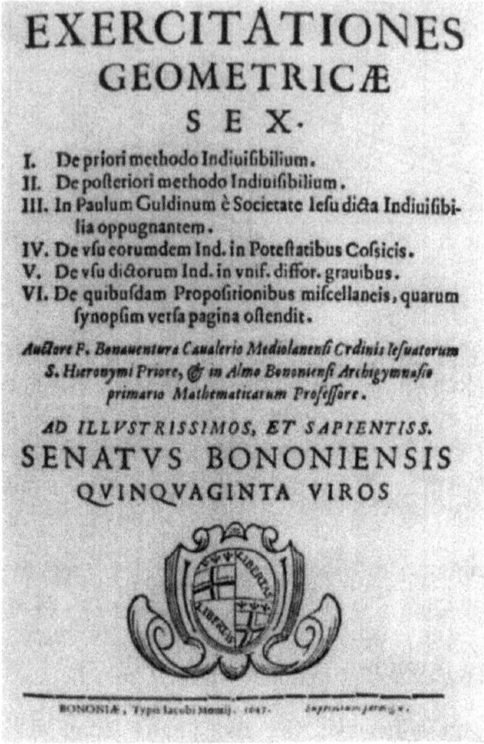

**EXERCITATIONES
GEOMETRICÆ
S E X·**

I.   De priori methodo Indiuisibilium.
II. De posteriori methodo Indiuisibilium.
III. In Paulum Guldinum è Societate Iesu dicta Indiuisibi-
     lia oppugnantem.
IV. De vsu eorumdem Ind. in Potestatibus Cossicis.
V.   De vsu dictorum Ind. in vnif. diffor. grauibus.
VI. De quibusdam Propositionibus miscellaneis, quarum
     synopsim versa pagina ostendit.

*Auctore F. Bonauentura Caualerio Mediolanensi Crdinis Iesuatorum
S. Hieronymi Priore, & in Almo Bononiensi Archigymnasio
primario Mathematicarum Professore.*

AD ILLVSTRISSIMOS, ET SAPIENTISS.
SENATVS BONONIENSIS
QVINQVAGINTA VIROS

BONONIÆ, Typis Iacobi Montij. 1647.   *Superiorum permissu.*

PHOTO 3.4. *Six Geometrical Exercises.*

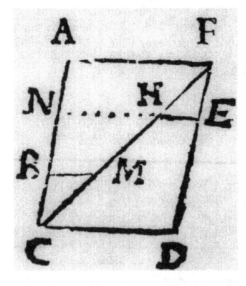

FIGURE 3.7. Cavalieri's theorem.

triangle *FDC* (because their altitude is the same, namely a.l. *AD*), and this ratio is 3. Hence a.c. of *AD* are equal to three times the product of a.l. of *AD* by a.s. of triangle *FDC*, and this is equal to the product of a.l. of triangle *ACF* by a.s. of triangle *FDC* plus the product of a.l. of triangle *FDC* by a.s. of the same triangle *FDC*, and this is equal to a.c. of triangle *FDC*. Hence a.c. of *AD* will be three times the sum of a.c. of triangle *FDC* and the product of a.l. of triangle *ACF* by a.s. of triangle *FDC*.

If now we resolve a.c. of *AD* into its parts, then we shall get a.c. of *ACF* plus a.c. of *FDC* plus three times the product of a.l. of triangle *FDC* by a.s. of triangle *ACF* plus three times the product of a.l. of triangle *ACF* by a.s. of triangle *FDC*. But three times the product of a.l. of triangle *ACF* by a.s. of triangle *FDC* is three times the same product. If we take it away, three times of what we take away remains. So that a.c. of *ACF* plus a.c. of *FDC* and three times the product of a.l. of triangle *FDC* by a.s. of triangle *FAC* are three times a.c. of triangle *FDC*. Now a.c. of triangle *ACF* plus a.c. of triangle *FCD* are twice a.c. of triangle *FCD*, since a.c. of triangle *ACF* will be equal to a.c. of triangle *FCD*.

Hence three times the product of a.l. of triangle *ACF* by a.s. of triangle *FCD* together with three times the product of a.l. of triangle *FCD* by a.s. of triangle *FAC* and a.c. of the triangles *ACF*, *FDC*, that is a.c. of parallelogram *AD*, are equal to the quadruple of a.c. of triangle *FDC* (of triangle *FAC*).

This is so, since the product of a.l. of *ACF* by a.s. of *FDC* is equal to the product of a.l. of *FDC* by a.s. of *ACF*, and this is so because of the equality of the lines and their squares in those triangles *FDC*, *ACF* which alternately correspond. Hence three times the product of a.l. of *ACF* with a.s. of *FDC* are equal to three times the product of a.l. of *FDC* and a.s. of *ACF*. This makes the proof clear.

Saving our mathematical commentary for a moment, let us see how Cavalieri quickly connects this to the areas bounded by higher parabolic curves ($py = x^n$ in modern notation) [166, p. 217].

Proposition 23. In any parallelogram such as *BD* [Figure 3.8] with base *CD* we draw an arbitrary parallel *EF* to *CD* and the diagonal *AC*, intersecting *EF* in *G*. Then $DA : AF = (CD$ or $EF) : FG$. We call *AC* the prime diagonal. Then we construct point *H* on *EF* such that $DA^2 : AF^2 = EF : FH$, and so on all parallels to *CD*, so that all lines like this *HF* end on a curve *CHA*. In a similar way we construct a curve *CIA*, where $DA^3 : AF^3 = EF : FI$, a curve *CLA* such that $DA^4 : AF^4 = EF : FL$, etc. We call *CHA* the second diagonal, *CIA* the third, *CLA* the fourth, etc., and similarly *AGCD* the first diagonal space of parallelogram *BD*, the trilinear figure *AHCD* the second, *AICD* the third, *ALCD* the fourth, etc. Then I say that parallelogram *BD* is twice the first, three times the second, four times the third, five times the fourth space, etc.

Notice that Cavalieri's results, like Archimedes', still consist of comparing areas and volumes via ratios, rather than asserting actual numerical values for them. A change in viewpoint would eventually be needed for the problem to become more tractable [21].

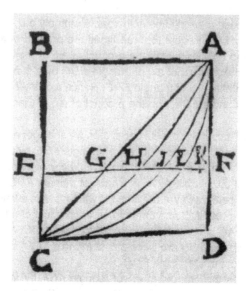

FIGURE 3.8. Cavalieri's curves.

Returning first to the proof of Proposition 21, the reader will wish to con-
vert Cavalieri's verbal and geometric algebra into modern algebraic language
(Exercise 3.20). Despite the emergence of considerable algebraic notation in the
Renaissance, Cavalieri's text is still remarkably verbal in nature.

The first paragraph of the proof simply uses a familiar algebraic identity. The
next claim, which ends by asserting that a certain ratio is three, relies on Cavalieri's earlier proof that the sum of all squares of AD is three times all squares
of triangle FDC, equivalent to the quadrature of the parabola already proven by
Archimedes. Cavalieri's Proposition 21 was the first higher parabola quadrature
after Archimedes.

Cavalieri goes on in Proposition 22 to prove the analogous result for fourth powers, namely that all quadruples of AD are the quintuple of all quadruples of FDC,
and he then states the general pattern. One can imagine how much more involved
and difficult even the next case was, and thus one can appreciate that people would
seek simpler methods of calculation. Cavalieri's Proposition 23 now provides a
geometric interpretation of these results as areas for the higher parabolic curves.
(How do his results on powers of segments in the parallelogram metamorphose
into his claims about the higher parabolas?)

Cavalieri's results were almost simultaneously found and demonstrated by others, like Fermat and Roberval, but by somewhat different means. All of them,
however, were elaborate and difficult methods to carry further. A few decades later
the breakthroughs of Leibniz and Newton, generally considered the founders of
the calculus, would provide general methods of calculation for these and many
other problems.

Exercise 3.20: Rewrite Cavalieri's proof of Proposition 21 using modern algebraic
language.

Exercise 3.21: Use Cavalieri's Principle to compute the volume of a sphere. Inscribe the sphere in a cylinder. Remove conical hollows as wide as possible from each end of the cylinder, meeting in the center. Show that the remaining volume is that of the sphere, according to Cavalieri's Principle.

Exercise 3.22: Use Cavalieri's Principle to find the volume of a segment of a sphere, obtained by slicing a plane through the sphere.

Exercise 3.23: Imagine the segment of Exercise 3.22 as the top of a filled ice-cream cone with tip at the center of the sphere. Find this volume also, and by comparing areas and volumes find the surface area of the curved spherical surface on top of the ice-cream cone. You may use knowledge of the surface area of the entire sphere, which Archimedes had determined.

Exercise 3.24: Imagine boring a round hole through the center of a sphere, leaving a spherical ring. Use the result of Exercise 3.23 to find the volume of the ring. Hint: Amazingly, your answer should depend only on the height of the ring, not the size of the original sphere. Also obtain your result directly from Cavalieri's Principle by comparing the ring with a sphere of diameter the height of the ring.

3.5   Leibniz's Fundamental Theorem of Calculus

Gottfried Wilhelm Leibniz and Isaac Newton were geniuses who lived quite different lives and invented quite different versions of the infinitesimal calculus, each to suit his own interests and purposes.

Newton discovered his fundamental ideas in 1664–1666, while a student at Cambridge University. During a good part of these years the University was closed due to the plague, and Newton worked at his family home in Woolsthorpe, Lincolnshire. However, his ideas were not published until 1687. Leibniz, in France and Germany, on the other hand, began his own breakthroughs in 1675, publishing in 1684. The importance of publication is illustrated by the fact that scientific communication was still sufficiently uncoordinated that it was possible for the work of Newton and Leibniz to proceed independently for many years without reciprocal knowledge and input. Disputes about the priority of their discoveries raged for centuries, fed by nationalistic tendencies in England and Germany.

Leibniz was born and schooled in Leipzig, studying law at the university there. Although he loved mathematics, he received relatively little formal encouragement. Later, after completing his doctorate at Altdorf, the university town of Nürenberg, he declined a professorship there, considering universities "monkish" places with learning, but little common sense, engaged mostly in empty trivialities. Instead, Leibniz entered public life in the service of princes, electors, and dukes, whom he served in legal and diplomatic realms, and in genealogical research trying to prove their royal claims. His work provided great opportunities for travel, and he interacted personally and by correspondence with philosophers and scientists throughout Europe, pursuing mathematics, the sciences, history, philosophy, logic, theology, and metaphysics. He was truly a genius of universal interests and contri-

PHOTO 3.5. Leibniz.

butions, leading him to concentrate on methodological questions, and to embark
on a lifelong project to reduce all knowledge and reasoning to a "universal characteristic." Although today we recognize his contributions to be of outstanding
importance, he died essentially neglected, and only his secretary attended his burial
[42, 157].

In 1672 Leibniz was sent to Paris on a diplomatic mission, beginning a crucially
formative four-year period there. Christiaan Huygens (1629–1695), from Holland, then the leading mathematician and natural philosopher in Europe, guided
Leibniz in educating himself in higher mathematics, and Leibniz's progress was
extraordinary.

Leibniz's discovery of the calculus emerged from at least three important interests [8, 21, 77]. First, as a philosopher his main goal was a general symbolic
language, enabling all processes of reason and argument to be written in symbols
and formulas obeying certain rules. His mathematical investigations were thus
merely part of a truly grand plan, and this explains his focus on developing useful
new notation and theoretical methods, rather than specific results. Indeed, it is his
notation and language for the calculus that we use today, rather than Newton's. He
sought and found a "calculus" for infinitesimal geometry based on new symbols
and rules.

FIGURE 3.9. Triangular numbers.

Second, Leibniz studied the relationship between difference sequences and sums, and then an infinitesimal version helped suggest to him the essential features of the calculus. This can be illustrated via a concrete problem that Huygens gave Leibniz in 1672: Consider the "triangular numbers" 1, 3, 6, 10, 15,..., the numbers of dots in triangular arrangements (Figure 3.9). These also occur in "Pascal's triangle" of binomial coefficients, and their successive differences are 2, 3, 4, 5,.... The triangular numbers are given by the formula $i(i+1)/2$ for $i = 1, 2,...$ (Can you verify this? Hint: Successive differences.) Huygens challenged Leibniz to calculate the infinite sum of their reciprocals,

$$\frac{1}{1} + \frac{1}{3} + \frac{1}{6} + \frac{1}{10} + \cdots + \frac{2}{n(n+1)} + \cdots,$$

and Leibniz proceeded as follows [77, pp. 60–61]. Each term $2/(i(i+1))$ in the sum equals the difference $2/i - 2/(i+1)$; i.e., the sum can be rewritten as

$$\left(\frac{2}{1} - \frac{2}{2}\right) + \left(\frac{2}{2} - \frac{2}{3}\right) + \left(\frac{2}{3} - \frac{2}{4}\right) + \left(\frac{2}{4} - \frac{2}{5}\right) + \cdots + \left(\frac{2}{n} - \frac{2}{n+1}\right) + \cdots.$$

The terms here can be regrouped and mostly canceled, so the partial sum of the first n original terms, displayed as a sum of differences, collapses, leaving $2 - 2/(n+1)$ as its sum, which leads to 2 as the sum of the original terms all the way to infinity. Viewed more generally, since the terms of the original series $2/(i(i+1))$ were recognized as being the successive differences of the terms in a new pattern (namely $2/i$), the nth partial sum of the original series can be computed, via the collapsing trick, as simply the difference between the first and nth terms in the new pattern. This observation expresses, in a discrete, rather than continuous, way, the essence of the Fundamental Theorem of Calculus, and Leibniz slowly came to realize this.

Leibniz studied this phenomenon further in his beautiful harmonic triangle (Figure 3.10 and Exercise 3.25), making him acutely aware that forming difference sequences and sums of sequences are mutually inverse operations. He used an analogy to think of the problem of area as a summation of infinitesimal differences, leading him to the connection between area and tangent.

The third crucial thread contributing to Leibniz's creation of the calculus was his conception of a "characteristic triangle" with infinitesimal sides at each point along a curve (see *GLC* or *(C)EC* in Figure 3.11). The two legs of the right triangle represent infinitesimal elements of change (successive differences) in the horizontal and vertical coordinates between the chosen point and an infinitesimally nearby point along the curve, and their ratio is thus the slope of the tangent line to the curve at the point. Leibniz wrote that these phenomena all came together as

$$1$$

$$\frac{1}{2} \qquad \frac{1}{2}$$

$$\frac{1}{3} \qquad \frac{1}{6} \qquad \frac{1}{3}$$

$$\frac{1}{4} \qquad \frac{1}{12} \qquad \frac{1}{12} \qquad \frac{1}{4}$$

$$\frac{1}{5} \qquad \frac{1}{20} \qquad \frac{1}{30} \qquad \frac{1}{20} \qquad \frac{1}{5}$$

$$\frac{1}{6} \qquad \frac{1}{30} \qquad \frac{1}{60} \qquad \frac{1}{60} \qquad \frac{1}{30} \qquad \frac{1}{6}$$

$$\frac{1}{7} \qquad \frac{1}{42} \qquad \frac{1}{105} \qquad \frac{1}{140} \qquad \frac{1}{105} \qquad \frac{1}{42} \qquad \frac{1}{7}$$

FIGURE 3.10. Leibniz's harmonic triangle.

"a great light" bursting upon him when he was studying Pascal's *Treatise on the Sines of a Quadrant of a Circle* [21, p. 203].

Combining Leibniz's connection between sums and successive differences with his connection between infinitesimal differences and tangent lines, we can begin to see a possible connection between area and tangent problems. Several other mathematicians had already been developing methods for finding tangent lines to curves, providing stimulus to Leibniz's ideas. Pierre de Fermat, for instance, illustrated his approach based on infinitesimals by calculating the tangent line to a parabola (Exercise 3.26).

In his early work with characteristic triangles and their infinitesimal sides, Leibniz derived relationships between areas that we today recognize as important general calculation tools (e.g., "integration by parts"), and while studying the quadrature of the circle, he discovered a strikingly beautiful result about an infinite sum, today named Leibniz's series:

$$1 - \frac{1}{3} + \frac{1}{5} - \frac{1}{7} + \cdots = \frac{\pi}{4}.$$

He began introducing and refining powerful notation for his ideas of sums and differences involving infinitesimals, ultimately settling on $d(x)$ for the infinitesimal differences between values of x, and on $\int$ (an elongated form of "s," as in Latin "summa") for the sum, or "integration," of infinitesimals. This $\int$ is analogous to Cavalieri's "all lines," so $\int y\,dx$ denotes the summation of the areas of all rectangles with length y and infinitesimal width dx. (Can you see why $d(xy) = x\,dy + y\,dx$? Hint: Think, like Leibniz, of x and y as successive partial sums, with dx and dy the differences between successive sums.)

Using the calculus he developed with these new symbols, Leibniz easily rederived many earlier results, such as Cavalieri's quadrature of the higher parabolas, and put in place the initial concepts, calculational tools, and notation for the enormous modern subject of analysis.

Although many of these seminal ideas are in Leibniz's manuscripts of 1675–1677, publication was slow. We will examine how he brought all these ideas together in his resolution of the problem of quadratures, in a 1693 paper *Supplementum geometriae dimensoriae, seu generalissima omnium tetragonismorum effectio per motum: similiterque multiplex constructio lineae ex data tangentium conditione* (More on geometric measurement, or most generally of all practicing of quadrilateralization through motion: likewise many ways to construct a curve from a given condition on its tangents) published in the first scientific journal *Acta Eruditorum* [110, pp. 294–301], [166, pp. 282–284]. Today we recognize his result as of such paramount importance that we call it the Fundamental Theorem of Calculus.

Leibniz, from

More on geometric measurement,
or most generally of all practicing of quadrilateralization
through motion: likewise many ways to construct a curve
from a given condition on its tangents.

I shall now show that the general problem of quadratures can be reduced to the finding of a line that has a given law of tangency (declivitas), that is, for which the sides of the characteristic triangle have a given mutual relation. Then I shall show how this line can be described by a motion that I have invented. For this purpose [Figure 3.11] I assume for every curve $C(C')$ a double characteristic triangle,[9] one, TBC, that is assignable, and one, GLC, that is inassignable, and these two are

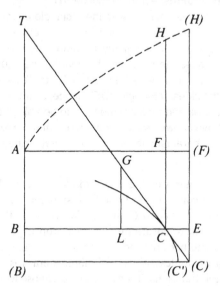

FIGURE 3.11. Leibniz's Fundamental Theorem of Calculus.

---

[9]In the figure Leibniz assigns the symbol (C) to two points, which we denote by (C) and (C'). If, with Leibniz, we write $CF = x$, $BC = y$, $HF = z$, then $E(C) = dx$, $CE = F(F) =$

similar. The inassignable triangle consists of the parts GL, LC, with the elements of the coordinates CF, CB as sides, and GC, the element of arc, as the base or hypotenuse. But the assignable triangle TBC consists of the axis, the ordinate, and the tangent, and therefore contains the angle between the direction of the curve (or its tangent) and the axis or base, that is, the inclination of the curve at the given point C. Now let $F(H)$, the region of which the area has to be squared, be enclosed between the curve $H(H)$, the parallel lines FH and $(F)(H)$, and the axis $F(F)$; on that axis let A be a fixed point, and let a line AB, the conjugate axis, be drawn through A perpendicular to AF. We assume that point C lies on HF (continued if necessary); this gives a new curve $C(C')$ with the property that, if from point C to the conjugate axis AB (continued if necessary) both its ordinate CB (equal to AF) and tangent CT are drawn, the part TB of the axis between them is to BC as HF to a constant [segment] a, or a times BT is equal to the rectangle AFH (circumscribed about the trilinear figure $AFHA$). This being established, I claim that the rectangle on a and $E(C)$ (we must discriminate between the ordinates FC and $(F)(C)$ of the curve) is equal to the region $F(H)$. When therefore I continue line $H(H)$ to A, the trilinear figure $AFHA$ of the figure to be squared is equal to the rectangle with the constant a and the ordinate FC of the squaring curve as sides. This follows immediately from our calculus. Let $AF = y$, $FH = z$, $BT = t$, and $FC = x$; then $t = zy : a$, according to our assumption; on the other hand, $t = y\, dx : dy$ because of the property of the tangents expressed in our calculus. Hence $a\, dx = z\, dy$ and therefore $ax = \int z\, dy = AFHA$. Hence the curve $C(C')$ is the quadratrix with respect to the curve $H(H)$, while the ordinate FC of $C(C')$, multiplied by the constant a, makes the rectangle equal to the area, or the sum of the ordinates $H(H)$ corresponding to the corresponding abscissas AF. Therefore, since $BT : AF = FH : a$ (by assumption), and the relation of this FH to AF (which expresses the nature of the figure to be squared) is given, the relation of BT to FH or to BC , as well as that of BT to TC, will be given, that is, the relation between the sides of triangle TBC. Hence, all that is needed to be able to perform the quadratures and measurements is to be able to describe the curve $C(C')$ (which, as we have shown, is the quadratrix), when the relation between the sides of the assignable characteristic triangle TBC (that is, the law of inclination of the curve) is given.

We make only a few remarks on the details of the text, confident that the reader can fill in the necessary connections. First, "inassignable" refers to infinitesimal characteristic triangles, such as GLC and $(C)EC$. Second, the mysterious constant segment a is present to ensure dimensional propriety; i.e., an area should be equated only with another area, not the length of a line, and a ratio of two lengths should be dimensionless. Even in Leibniz's era this view was still carried from ancient Greek traditions. Today we could choose to view a as simply a choice of unit length for measurement. (From this point of view, why does a not

---

dy, and $H(H)(F)F = z\, dy$. First Leibniz introduces curve $C(C')$ with its characteristic triangle, and then later reintroduces it as the squareing curve [*curva quadratrix*] of curve $AH(H)$.

really affect the final answer?) Third, note that Leibniz is not using the Cartesian coordinates pioneered earlier in his century quite as we do, but he is nevertheless measuring all locations along two perpendicular axes from a common origin A. Interestingly, he has no qualms about depicting the vertical coordinate for the quadratrix as increasing downwards, while for the curve to be squared it increases upwards (Why do you think he does this?). Finally, although Leibniz shows the quadratrix (squaring curve) only near the point C, in fact it must originate at A (Why?).

Leibniz never explains exactly what the meaning of his inassignables is, and on this he vacillated in his writings. To him, the most important criterion was that his rules for applying the new language worked, and he stated that applying them as if they were the rules of algebra would dispense with the "necessity of imagination" [21, p. 208].

Indeed, while the Fundamental Theorem of Calculus does not actually dispense with the need for imagination, it reduces every quadrature problem to finding a curve with a "given law of tangency," i.e., an inverse tangent problem.

While Fermat and others had shown that finding the tangent to a given curve is often possible, the inverse problem of finding a curve, given only its law of tangency, is generally much harder. It is not always possible to accomplish algebraically, even when the law of tangency is given algebraically.

Nonetheless, let us see some examples of how the Fundamental Theorem can be applied. Suppose (adhering to Leibniz's choices for coordinates, x and y) that we wish to square the curve $x = y^2$ (quadrature of the parabola again). According to his theorem, we must find a curve with y^2 as its law of tangency. This essentially involves guessing an answer based on experience, $y^3/3$ in this case, and then verifying that it works.

Let us imagine how Leibniz might have done this calculation. The law of tangency for $y^3/3$ will be obtained from its infinitesimal (inassignable) characteristic triangle as the ratio of the respective increments in x and y, i.e., $dx : dy$. Since dy is the increment (difference) between successive values (of the sums) y and $y+dy$, with corresponding values $x = y^3/3$ and $x + dx = (y + dy)^3/3$, we calculate

$$dx : dy = \frac{dx}{dy} = \frac{(x+dx) - x}{dy} = \frac{(y+dy)^3 - y^3}{3 \cdot dy}$$

$$= \frac{y^3 + 3y^2 \cdot dy + 3y \cdot (dy)^2 + (dy)^3 - y^3}{3 \cdot dy}$$

$$= y^2 + y \cdot dy + \frac{(dy)^2}{3} = y^2,$$

as claimed. Leibniz explains that the final equality holds by dropping remaining infinitesimal terms or, going back one step, because $(dy)^2$ is infinitesimally small in comparison with dy.

Thus, according to the Fundamental Theorem of Calculus, the area under the parabola but above the y-axis, and from the origin to a specific value y for the ordinate, is given by $y^3/3$. Other examples are in the exercises.

Both Leibniz and Newton had their calculus attacked by others for their use of infinitesimals. One of the most eloquent and stinging criticisms came from Bishop George Berkeley's (1685–1753) polemic comparing the validity of science and mathematics with that of religion, entitled *The Analyst: Or a Discourse Addressed to an Infidel Mathematician. Wherein It Is Examined Whether the Object, Principles, and Inferences of the Modern Analysis Are More Distinctly Conceived, or More Evidently Deduced, than Religious Mysteries and Points of Faith. "First Cast the Beam Out of Thine Own Eye; and Then Shalt Thou See Clearly to Cast Out the Mote Out of Thy Brother's Eye,"* addressed to Edmund Halley (discoverer of the comet), a friend and defender of Newton in the early eighteenth century [21, pp. 224–225].

Berkeley writes:

Whereas then it is supposed that you apprehend more distinctly, consider more closely, infer more justly, and conclude more accurately than other men, and that you are therefore less religious because more judicious, I shall claim the privilege of a Freethinker; and take the liberty to inquire into the object, principles, and method of demonstration admitted by the mathematicians of the present age, with the same freedom that you presume to treat the principles and mysteries of Religion [160, pp. 627–634].

Berkeley then proceeds to ridicule the ideas and terminology of the calculus, which involved not only infinitesimals like Leibniz's dx and dy above, but infinitesimal differences of these differences (i.e., second differences), etc. Berkeley concludes therefore:

All these points, I say, are supposed and believed by certain rigorous exactors of evidence in religion, men who pretend to believe no further than they can see. That men who have been conversant only about clear points should with difficulty admit obscure ones might not seem altogether unaccountable. But he who can digest...a second or third difference, need not, methinks, be squeamish about any point of divinity...

Berkeley explicitly rips apart the type of tangent calculation we have just seen for $x = y^3$. He says of the differences (increments) dx and dy:

For when it is said, let the increments vanish, i.e., let the increments be nothing, or let there be no increments, the former supposition that the increments were something, or that there were increments, is destroyed, and yet a consequence of that supposition, i.e., an expression got by virtue thereof, is retained. Which...is a false way of reasoning. Certainly when we suppose the increments to vanish, we must suppose their proportions, their expressions, and everything else derived from the supposition of their existence, to vanish with them....

Berkeley admits that the calculus produces correct answers, but for no solid reasons:

I have no controversy about your conclusions, but only about your logic and method: how you demonstrate? what objects you are conversant with, and whether you conceive them clearly? what principles you proceed upon; how sound they may be; and how you apply them? ...

Now, I observe, in the first place, that the conclusion comes out right, not because the rejected square of dy was infinitely small, but because this error was compensated for by another contrary and equal error....

And what are these...evanescent increments? They are neither finite quantities, nor quantities infinitely small nor yet nothing. May we not call them the ghosts of departed quantities?

And finally, in one of a series of "questions" issued as a challenge to mathematicians who criticize religion:

Question 64. Whether mathematicians, who are so delicate in religious points, are strictly scrupulous in their own science? Whether they do not submit to authority, take things upon trust, and believe in points inconceivable? Whether they have not *their* mysteries, and what is more, their repugnances and contradictions?

While Berkeley's mathematical criticisms were largely valid, it is clear he had primarily a religious axe to grind. By arguing that their calculus was no more scientific than theology, and that it too was also built only on faith, he wanted to shame mathematicians into refraining from criticizing religion. Berkeley admitted that the calculus led to correct answers, and claimed that this resulted from a "compensation of errors," in which the multiple errors implicit in the rules of calculus somehow cancel each other's effects, thus arriving "though not at Science, yet at Truth, For Science it cannot be called, when you proceed blindfold, and arrive at the Truth not knowing how or by what means" [77, pp. 88–89]. We will see how some of his mathematical criticisms slowly began to be resolved.

Exercise 3.25: Study Leibniz's harmonic triangle of successive differences. Determine formulas for all the terms, and show how to find the sums of various infinite series of successive differences.

Exercise 3.26: Read and explain Fermat's method of finding the tangent line to a parabola via infinitesimals. Compare the notation of the French and English translations [59, III, pp. 121–23][58, pp. 358–359][166, pp. 223–24] with the Latin original [59, I, pp. 133–35]. Is Fermat's result already in Proposition 2 from Archimedes' *Quadrature of the Parabola* [3]?

Exercise 3.27: Verify that our quadrature of the parabola with Leibniz's Fundamental Theorem of Calculus effectively yields the same result as those of Archimedes and Cavalieri.

Exercise 3.28: Calculate the area under $x = y^3$, and in general $x = y^n$, using Leibniz's Fundamental Theorem of Calculus and his notation.

Exercise 3.29: Can you extend the results of Exercise 3.28 to exponents n other than 1, 2, 3,..., e.g., fractional or negative powers?

Exercise 3.30: Can you calculate $d\left(\frac{y}{x}\right)$ as Leibniz might have?

Exercise 3.31: Can you calculate $d(\sin(x))$ or $d(10^x)$ or $d(e^x)$ à la Leibniz? Can you use these to find areas?

3.6   Cauchy's Rigorization of Calculus

Augustin-Louis Cauchy was born in 1789, the year the French Revolution began. He was the eldest of six children, and his father, a student of classics and barrister in Normandy, ensured him an excellent education, leading at age 16 to admittance at the elite Ecole Polytechnique. He subsequently entered the Ecole des Ponts et Chaussées (College of Bridges and Roadways), became an engineer, and worked in 1810 at Cherbourg Harbor, where Napoleon was building a naval base for the invasion of England.

By 1816 Cauchy had established himself as an important mathematical researcher, and his winning essay for a French Academy of Sciences prize contest on the propagation of waves on a liquid surface is today a classic of hydrodynamics. He had been appointed professor at the Ecole Polytechnique, and also been appointed, not elected, to the Academy of Sciences while others were being expelled in the political turmoil between royalty and bourgeoisie. Cauchy was an ultraconservative Royalist and Catholic, and was accused of being bigoted, selfish, narrow-minded, and naive. It is no surprise that his own fortunes swept back and forth with politics and religion of the period. For instance, when the 1830 revolution replaced a Bourbon king with a king of the bourgeoisie, he refused to take the oath of allegiance, and went into self-imposed exile for 8 years.

Cauchy was a prolific mathematician who opened broad new vistas in many areas of mathematics, such as differential equations, elasticity theory, and functions of a complex variable, for which he is probably best known to mathematicians today. He was one of the great mathematical powers of the early nineteenth century [42].

In the late eighteenth century it had become ever clearer that the foundations of calculus needed serious attention, and the Berlin Academy of Sciences, urged especially by Joseph-Louis Lagrange (1736–1813), one of the leading mathematicians of the time, made this the topic of a prize problem in 1784. The scope of concern was large, indicated by the statement that the prize was to be awarded for "a clear and precise theory of what is called *Infinity* in mathematics" [72, p. 41].

None of the solutions submitted was even close to satisfactory. Lagrange's own subsequent treatise on the subject, *Fonctions Analytiques* (Analytic Functions), attempted to found calculus entirely on the algebra of infinite series, proposing that every function should be represented by a power series (see the Introduction and Appendix), whose coefficients hold all the data of calculus, e.g., the derivative and antiderivative.

PHOTO 3.6. Cauchy.

Unfortunately, this bold idea failed in the end to capture adequately the nature of the functions studied via calculus: There were important functions for calculus that could not be represented as power series, and moreover, two different functions could have the same power series (Exercise 3.39), as Cauchy would say, politely but firmly rejecting Lagrange's framework when he introduced his own [72, p. 54].

Cauchy's fresh development of the fundamental features of calculus was published in now classic texts based on lectures at the Ecole. His first step towards a rigorous theory was *Cours d'analyse* (A Course in Analysis). In this 1821 volume, whose subtitle *Analyse Algébrique* (Algebraic Analysis) suggests that Cauchy was trying to meld his new framework with Lagrange's dream of a purely algebraic calculus, he presented a theory of limits, continuous functions, and infinite series. Cauchy's presentation of the calculus came two years later in his *Résumé des Leçons Données a L'Ecole Royale Polytechnique sur le Calculus Infinitésimal* (Compendium of Lectures Presented on the Infinitesimal Calculus at the Royal Polytechnic College) [30, vol. 4, pp. 5–261], [31]. He presents first the differential calculus in twenty brief lessons, and then the integral calculus in twenty more. Cauchy covers essentially everything seen today in a year-long calculus course, including some calculus of functions of several variables, and even topics using imaginary numbers. His treatment finally put to rest objections like Bishop Berkeley's tirade against the calculus of Newton and Leibniz.

Cauchy's exposition makes enticing reading, and we recommend it highly. Our selections portray the solution of the area problem within this advancing framework of modern analysis. Many of our exercises are calculations Cauchy presents in his lectures.

In the introduction to *Cours d'analyse, Analyse Algébrique*, Cauchy wrote "As for methods, I have sought to give them all the rigor that we require in geometry, so as never to appeal to arguments drawn from the generality of algebra" [77, p. 96]. Here he rejects the then still current practice of assuming that algebraic facts valid in one setting will also hold in a setting of greater generality; for instance, Leibniz's faith that algebraic results about ordinary geometrical magnitudes generalize to infinitesimals, or Lagrange's faith in the algebra of infinite power series (Exercise 3.46).

In the introduction to *Leçons...sur le Calculus Infinitésimal*, Cauchy writes:

> My principal goal has been to reconcile the rigor, of which I made a rule in my *Cours d'analyse*, with the simplicity which results from direct consideration of infinitely small quantities.... In the integral calculus, it seemed necessary to me to demonstrate in general the existence of *integrals* or *primitive functions* before indicating their various properties. For this it was first necessary to establish the notion of *integrals taken between given limits* or *definite integrals*.

Here Cauchy explains that he intends to avoid the earlier circularity between derivatives and integrals by defining the two completely separately, and then linking them by the Fundamental Theorem of Calculus. Thus Cauchy will for the first time really define what is meant by area, instead of taking it for granted as having intuitive meaning. Cauchy bases his new analysis of functions on the theory of limits and arithmetic of real numbers, freeing it from both geometry and freewheeling algebraic generalizations, while building in rigor of the standard of geometry. In concert with this we find not a single diagram in the entire book! Much of his approach is actually based on approximation methods of algebra and analysis developed in the eighteenth century, using inequalities between real numbers as the foundation for the calculus.

Augustin-Louis Cauchy, from
Lectures on the Infinitesimal Calculus

FIRST LESSON
Variables, their Limits, and infinitely small Quantities

One calls a *variable* that which one considers as receiving successively various different values.... When the values successively attributed to the same variable approach a fixed value indefinitely, in such a manner as to end up differing from it as little as one could wish, this latter value is called the *limit* of all the others. So, for example, the area of a circle is the limit to which the areas of inscribed regular polygons converge as the number of sides grows more and more; and the radius vector drawn from the center of a hyperbola to a point on the curve which moves farther and farther from the center forms an angle with the x-axis which has as its limit the angle formed by the asymptote with this axis; &c.... We indicate the limit to which a variable converges by the abbreviation *lim.* placed before the variable.

In the earlier *Algebraic Analysis*, Cauchy gave an additional example, that

an irrational number is the limit of the various fractions which provide values that approximate it more and more closely [58, p. 566].

This last assertion illustrates an aspect of Cauchy's rigorization that needed further clarification, since one really needs a proper definition of what irrational numbers are before one can start giving examples of their use and features; Cauchy was not able to do this, and it would fall to future generations to provide this understanding.

Cauchy continues in Lesson 1 to analyze two very important limits, of the "variable expressions" $\frac{\sin \alpha}{\alpha}$ and $(1 + \alpha)^{\frac{1}{\alpha}}$ as α converges to zero (Exercises 3.32 and 3.33).

Near the end of the lesson he uses the limit concept to define infinitesimals (*not* vice versa, as had his predecessors):

When the successive numerical values of the same variable decrease indefinitely in such a manner as to fall below any given number, this variable becomes what one calls an *infinitesimal* or an infinitely small quantity. A variable of this kind has zero for its limit. Such is the variable α in the preceding calculations.

SECOND LESSON
Continuous and discontinuous Functions.
Geometric representation of continuous Functions

When the function $f(x)$ has a unique and finite value for all values of x between two given limits, and the difference

$$f(x + i) - f(x)$$

is always an infinitesimal quantity between these limits, one says that $f(x)$ is a *continuous function* of the variable x between these limits. One says also that the function $f(x)$ is a continuous function of this variable x in the neighborhood of a particular value of x, whenever it is continuous between two limits, even very close, enclosing the value in question.

Here Cauchy appears to rephrase his definition of continuity, or does he actually give two different definitions of continuity, one for continuity of a function on an entire interval (global), the other for continuity at a single value x (local)? The delicate differences between these two situations exemplify the sort of lack of rigor that later haunted Cauchy and others; in the end even Cauchy's rigorization was only a step, albeit an enormous one, in a longer process.

Cauchy studies some examples (Exercise 3.34), and finally the connection between continuity and the geometry of curves:

Now let us imagine that one constructs the curve having the equation $y = f(x)$ in rectangular coordinates. If the function $f(x)$ is continuous between the limits $x = x_0, x = X$, there will be a unique ordinate corresponding to each abscissa x included between these limits; and moreover, when x increases by an infinitely small quantity Δx, y will increase by an infinitely small quantity Δy. Consequently, to two closely approaching abscissas $x, x + \Delta x$, there correspond two closely approaching points, because their distance $\sqrt{\Delta x^2 + \Delta y^2}$ will itself be an infinitely small quantity....

Thus the geometric content of continuity is that the graph of $f(x)$ has no jumps (discontinuities). It is tempting to assume that continuity includes the assumption of all intermediate values by the function $f(x)$, as we discussed in the Introduction; but this is an example of something Cauchy could not completely prove, since the fine arithmetic structure of the real numbers was still out of reach, and would remain so for another half-century.

THIRD LESSON
Derivatives of Functions of a single Variable

When the function $y = f(x)$ remains continuous between two given limits of the variable x, and one assigns to the variable a value between these limits, an infinitesimal increment of the variable produces an infinitesimal increment of the function itself. Consequently, if we then set $\Delta x = i$, the two terms of the *difference quotient*

(1)
$$\frac{\Delta y}{\Delta x} = \frac{f(x + i) - f(x)}{i}$$

will be infinitely small quantities. But whereas these terms simultaneously approach the limit zero, the ratio itself may converge to another limit, either positive or negative. This limit, when it exists, has a definite value for each particular value of x; but it varies with x. Thus, for example, if we take $f(x) = x^m$, m being a [positive] whole number, the ratio between the infinitesimal differences will be

$$\frac{(x + i)^m - x^m}{i} = mx^{m-1} + \frac{m(m - 1)}{1 \cdot 2}x^{m-2}i + \cdots + i^{m-1}$$

and it will have as limit the quantity mx^{m-1}, that is to say, a new function of the variable x. The same will hold in general; only the form of the new function which serves as the limit of the ratio $\frac{f(x+i)-f(x)}{i}$ will depend upon the form of the given function $y = f(x)$. To indicate this dependence, one gives the new function the name *derivative* and indicates it, using an accent, by the notation

$$y' \text{ or } f'(x).$$

Cauchy goes on to produce formulas for derivatives of many common functions, and also obtains the basic differentiation rules of algebraic calculation from his definition, as Leibniz had originally done (Exercises 3.35–3.38).

In the rest of the first twenty lessons Cauchy develops a full theory of differential calculus based on the derivative, including its use for finding maximum and minimum values of functions of one or multiple variables.

With Lesson 21 Cauchy turns to integral calculus, and thus to the theme of the area problem, defining the integral quite separately from the derivative, unlike Leibniz, for whom the two were mysteriously intertwined. Cauchy's full development of the differential calculus separately and prior to the integral calculus was quite antihistorical, but persists to this day in teaching.

TWENTY-FIRST LESSON
Definite Integrals

Supposing that the function $y = f(x)$ is continuous with respect to the variable x between the two finite limits $x = x_0$, $x = X$, one designates by $x_1, x_2, \ldots,$ x_{n-1} new values of x interposed between these limits, and which are always either increasing or decreasing between the first limit and the second. One can use these values to divide the difference $X - x_0$ into elements

(1) $$x_1 - x_0, x_2 - x_1, x_3 - x_2, \ldots, X - x_{n-1},$$

which will all be the same sign. This done, imagine that we multiply each element by the value of $f(x)$ corresponding to the *origin* [left-hand end point] of that element: that is, the element $x_1 - x_0$ by $f(x_0)$, the element $x_2 - x_1$ by $f(x_1)$, &c., and finally the element $X - x_{n-1}$ by $f(x_{n-1})$; and let

(2) $$S = (x_1 - x_0)f(x_0) + (x_2 - x_1)f(x_1) + \cdots + (X - x_{n-1})f(x_{n-1})$$

be the sum of the products thus obtained. The quantity S clearly will depend upon, first, the number n of elements into which one has divided the difference $X - x_0$, and second, the values of these elements and therefore the mode of division adopted. Now it is significant to observe that if the numerical values of the elements become very small and the number n very large, the mode of division will have only an imperceptible influence on the value of S. This in fact can be proved as follows....

...and, if one makes the numerical values of these elements decrease indefinitely while their number increases, the value of S will ultimately become perceptibly constant, or, in other words, it ultimately attains a certain limit which will depend uniquely on the form of the function $f(x)$ and on the extreme values x_0, X of the variable x. This limit is what one calls a *definite integral*.

While Cauchy develops the definite integral independent of geometric considerations, clearly the sum S represents an approximation for the area under the graph of $y = f(x)$ between x_0 and X if $f(x)$ is always positive (Exercise 3.40). And then the integral itself yields the actual area, as Cauchy justifies in Lesson 23.

At the end of Lesson 21 Cauchy motivates and explains his notation for the definite integral:

Let us now observe that, if one denotes by $\Delta x = h = dx$ a finite increase in the variable x, then the different terms which form the value of S, such as the products

$(x_1 - x_0) f(x_0)$, $(x_2 - x_1) f(x_1)$, &c. will all be included in the general formula

(8) $h f(x) = f(x) dx$

from which one can deduce them one after the other by setting first $x = x_0$, and $h = x_1 - x_0$, then $x = x_1$ and $h = x_2 - x_1$, &c. One can thus state that the quantity S is a sum of products like the expression (8); this one can express using the character $\sum$ by writing

(9) $S = \sum h f(x) = \sum f(x) \Delta x.$

As for the definite integral to which the quantity S converges, as the elements of the difference $X - x_0$ become infinitely small, it is convenient to represent it by the notation $\int h f(x)$ or $\int f(x) dx$, in which the character $\int$ substituted for the character $\sum$ indicates, no longer a sum of products as in expression (8), but the limit of a sum of this type. Moreover, because the value of the definite integral under consideration depends on the extreme values of the variable x, it is convenient to place these two values, the first below, the second above the character $\int$, which one consequently designates by [the notation]

(10) $\int_{x_0}^{X} f(x) dx....$

[This] notation, conceived by M. *Fourier*, is the simplest.

In Lesson 22, Cauchy calculates the exact numerical values of several important definite integrals directly from the definition (Exercise 3.41), and also gives a numerical approximation with accuracy bound for $\int_0^1 dx/(1 + x^2)$ (Exercise 3.42).

After obtaining some important general properties of definite integrals, including in particular (7) of Lesson 23, that

$$\int_{x_0}^{X} f(x) dx = \int_{x_0}^{\xi} f(x) dx + \int_{\xi}^{X} f(x) dx, \text{ for } x_0 \leq \xi \leq X$$

(draw a picture to see that this is plausible), Cauchy is ready to prove the Fundamental Theorem of Calculus. It links his definite integral to differentiation, thus marrying the differential and integral calculus, and solves the area problem.

TWENTY-SIXTH LESSON
Indefinite Integrals

If in the definite integral $\int_{x_0}^{X} f(x) dx$ one varies one of the two limits of integration, for example X, the integral itself will vary with that quantity. And if the limit X, now become a variable, is replaced by x, we will obtain as a result a new function of x, which will be called an integral taken from the *origin* $x = x_0$. Let

(1) $\mathcal{F}(x) = \int_{x_0}^{x} f(x) dx$

be that new function. One obtains from formula (19) [Twenty-Second Lesson] that

(2) $\mathcal{F}(x) = (x - x_0) f[x_0 + \theta (x - x_0)], \mathcal{F}(x_0) = 0,$

θ being a [nonnegative] number less than [or equal to] one; and from formula (7) [Twenty-Third Lesson]

$$\int_{x_0}^{x+\alpha} f(x)\,dx - \int_{x_0}^{x} f(x)\,dx = \int_{x}^{x+\alpha} f(x)\,dx = \alpha f(x+\theta\alpha),$$

or

(3) $$\mathcal{F}(x+\alpha) - \mathcal{F}(x) = \alpha f(x+\theta\alpha).$$

It follows from equations (2) and (3) that if the function $f(x)$ is finite and continuous in the neighborhood of some particular value of the variable x, the new function $\mathcal{F}(x)$ will not only be finite but also continuous in the neighborhood of that value, since an infinitely small increment of x will correspond to an infinitely small increment in $\mathcal{F}(x)$. Thus if the function $f(x)$ remains finite and continuous from $x = x_0$ to $x = X$, the same will hold for the function $\mathcal{F}(x)$. Let us add that, if both members of formula (3) are divided by α, we may conclude, by passing to the limits, that

(4) $$\mathcal{F}'(x) = f(x).$$

Thus the integral (1) considered as a function of x, has as its derivative the function $f(x)$ contained under the sign $\int$ in this integral.

The equation Cauchy quotes from Lesson 22 is called the Mean Value Theorem for Integrals. Draw a picture to see whether it seems believable and why it is called this. We will not examine Cauchy's proof of it here, which relies on the Intermediate Value Theorem we discussed in the Introduction, and which thus remained as a point of rigor to be resolved later in the nineteenth century. Cauchy's main result here, equation (4), is today considered an alternative form of the Fundamental Theorem of Calculus, since it too indicates the inverse nature of the operations of differentiation and integration.

The Fundamental Theorem of Calculus as we saw it in Leibniz's text will now emerge from the relationship in (4) as Cauchy solves two problems (we slightly alter his notation in three spots to fit with that of our earlier selections).

First Problem. *One wants a function $\omega(x)$ whose derivative $\omega'(x)$ should be constantly null. In other words, one intends to solve the equation*

(6) $$\omega'(x) = 0.$$

Solution. If one wishes the function $\omega(x)$ to be finite and continuous between $x = -\infty$ and $x = +\infty$, then, designating by x_0 a particular value of the variable x, one derives from formula (6) [Seventh Lesson] $\omega(x) - \omega(x_0) = (x - x_0)\,\omega'[x_0 + \theta(x - x_0)]$ [θ being a nonnegative number less than or equal to one], and consequently

(7) $$\omega(x) = \omega(x_0),$$

or, denoting by c the constant quantity $\omega(x_0)$,

(8) $$\omega(x) = c.$$

Thus the function $\omega(x)$ must reduce to a constant, and maintains the same value c, between $x = -\infty$ and $x = +\infty$. One may add that this unique value will be entirely arbitrary, since formula (8) satisfies equation (6), whatever c may be.

In a nutshell: The only functions with zero as their derivative are the constant functions. The equation Cauchy quotes from Lesson 7 is called the Mean Value Theorem, a crucial result in the differential calculus (draw a graph to see whether you believe that this is plausible for any function with a derivative).

Second Problem. To find the general value of y suitable for satisfying the equation

$$(11) \qquad\qquad [y' = f(x)].$$

Solution. If one denotes by $F(x)$ a particular value of the unknown y, and by $F(x)+\omega(x)$ its general value, one deduces from formula (11), which both these values must satisfy, $F'(x) = f(x)$, $F'(x) + \omega'(x) = f(x)$, and consequently $\omega'(x) = 0$. On the other hand it follows from equation [(4)] that one satisfies formula (11) by taking $y = \int_{x_0}^{x} f(x)\,dx$. Thus the general value of y will be

$$(12) \qquad\qquad y = \int_{x_0}^{x} f(x)\,dx + \omega(x),$$

where $\omega(x)$ denotes a function suitable for satisfying equation (6). This general value of y, which includes, as a particular case, the integral (1), and which keeps the same form no matter what the origin x_0 is of this integral, is represented in calculation by the simple notation $\int f(x)\,dx$, and receives the name of *indefinite integral*....

If the function $F(x)$ differs from the integral (1), the general value of y, or $\int f(x)\,dx$, can always be presented in the form

$$(15) \qquad\qquad \int f(x)\,dx = F(x)+\omega(x),$$

and must reduce to the integral (1), for a particular value of $\omega(x)$ which satisfies both equation (6) and

$$(16) \qquad\qquad \mathcal{F}(x) = \int_{x_0}^{x} f(x)\,dx = F(x)+\omega(x).$$

If in addition the functions $f(x)$ and $\mathcal{F}(x)$ are both continuous between the limits $x = x_0$, $x = X$, then the function $\mathcal{F}(x)$ will itself be continuous and consequently $\omega(x) = \mathcal{F}(x)-F(x)$ retains the same constant value between these limits, between which one will have $\omega(x) = \omega(x_0)$, $\mathcal{F}(x) - F(x) = \mathcal{F}(x_0) - F(x_0) = -F(x_0)$, $\mathcal{F}(x) = F(x) - F(x_0)$,

$$(17) \qquad\qquad \int_{x_0}^{x} f(x)\,dx = F(x) - F(x_0).$$

Finally, if in equation (17) one sets $x = X$, one finds

$$(18) \qquad\qquad \int_{x_0}^{X} f(x)\,dx = F(X) - F(x_0).$$

Thus from equations (15), (17), and (18), if given a particular value $F(x)$ of y, satisfying formula (11), one can deduce, first, the value of the indefinite integral $\int f(x)\,dx$, and second, the values of the two definite integrals $\int_{x_0}^x f(x)\,dx$, $\int_{x_0}^X f(x)\,dx$, if the functions $f(x)$, $F(x)$ remain continuous between the limits of these two integrals.

Here we have seen the Fundamental Theorem of Calculus in slightly greater generality than Leibniz presented it, based on Cauchy's definition of the definite integral as a limit of sums. All arguments are founded on the analysis of functions, variables, and limits, using arithmetic rather than geometry. The theorem tells us that to compute a definite integral of f, we may seek any F whose derivative is f (F is then called an antiderivative for f), and take the difference in the values of F between the two endpoints. Cauchy's immediate example is the attractive result $\int_0^1 dx/\left(1 + x^2\right) = \pi/4$ (Exercise 3.43).

We should reiterate that the task of finding formulas for antiderivatives is neither easy nor always possible, but the Fundamental Theorem of Calculus is nonetheless the central tool of calculus and analysis. It achieved its modern formulation and proof with Cauchy, and enabled the easy computation of many areas and other quantitative phenomena.

In the Introduction we indicated that Riemann later broadened Cauchy's definition of the integral, and that by the twentieth century it had been replaced by a whole new theory, which encompassed highly discontinuous functions.

Riemann did not restrict himself to continuous functions, as did Cauchy, and he allowed Cauchy's sums

$$S = (x_1 - x_0)f(x_0) + (x_2 - x_1)f(x_1) + \cdots + (X - x_{n-1})f(x_{n-1}),$$

which were restricted to using the value of f at only the left endpoint of each subinterval, to be formed instead using the value of f at any point in the subinterval. Thus he considered sums

$$S = (x_1 - x_0)f(c_0) + (x_2 - x_1)f(c_1) + \cdots + (X - x_{n-1})f(c_{n-1}),$$

where $x_i \leq c_i \leq x_{i+1}$, and then required, as did Cauchy, that these sums should approach a fixed value as the subinterval lengths approach zero, which value is defined to be the integral of the function. With this definition Riemann managed to characterize exactly which functions have integrals, i.e., to what extent they can have discontinuities and still be integrable. While we will not describe his characterization, we shall consider an interesting example.

Dirichlet introduced a curious function, defined on the interval $0 \leq x \leq 1$ by letting $f(x) = 0$ if x is rational and $f(x) = 1$ if x is irrational. Now, since every interval contains both rational and irrational numbers, this function is discontinuous everywhere, and clearly the Riemann sum above, for any choice of subintervals, can be made to equal either zero or one with appropriate choices of the c_i. Thus there cannot possibly be a limit as the subinterval lengths decrease, so Dirichlet's function has no (Riemann) integral. Spurred by examples like Dirichlet's function, integration theory was revolutionized by work culminating in the new expanded theory of Henri Lebesgue at the turn of the twentieth century. In Lebesgue's theory, Dirichlet's function is integrable! Let us end by sketching roughly how.

Lebesgue's theory inverts the roles played by inputs and function values in defining the integral. The Cauchy–Riemann approach to the integral focuses first on intervals of inputs, then on the sizes of possible values of f on those intervals. Lebesgue, on the other hand, focuses first on possible function values, and then measures the length (more generally, the "measure") of the set of inputs giving those values. As Lebesgue himself expressed it most delightfully, in his article *The development of the integral concept*:

> The geometers of the seventeenth century considered the integral of $f(x)$...as the sum of an infinity of indivisibles, each of which was the ordinate, positive or negative, of $f(x)$. Very well! We have simply grouped together the indivisibles of comparable size. We have, as one says in algebra, collected similar terms. One could say that, according to Riemann's procedure, one tried to add the indivisibles by taking them in the order in which they were furnished by the variation of x, like an unsystematic merchant who counts coins and bills at random in the order in which they came to hand, while we operate like a methodical merchant who says:
>
> I have $m(E_1)$ pennies which are worth $1 \cdot m(E_1)$,
>
> I have $m(E_2)$ nickels worth $5 \cdot m(E_2)$,
>
> I have $m(E_3)$ dimes worth $10 \cdot m(E_3)$, etc.
>
> Altogether then I have
>
> $$S = 1 \cdot m(E_1) + 5 \cdot m(E_2) + 10 \cdot m(E_3) + \cdots.$$
>
> The two procedures will certainly lead the merchant to the same result because no matter how much money he has, there is only a finite number of coins and bills to count. But for us who must add an infinite number of indivisibles the difference between the two methods is of capital importance [106].

For Dirichlet's function there are only two values. Zero is the value on the set of rational numbers, one is the value on irrationals. In the spirit of Lebesgue's methodical merchant, we should then consider the sum

$$S = 0 \cdot m(Q) + 1 \cdot m(I),$$

where Q and I are the sets of rational and irrational numbers in the interval $[0, 1]$, and the function $m(\ )$ measures the generalized "length" of these sets. Of course, they are not intervals, and one of the greatest endeavors of the decades leading to Lebesgue's theory was the extension of the notion of geometric length to other types of sets than just intervals. Today we call this *measure theory*.

Here is an argument from measure theory for why $m(Q) = 0$! The set Q is countable (see the set theory chapter for details), i.e., we may label its elements $r_0, r_1, r_2, \ldots, r_i, \ldots$. Now, for any positive number ϵ, enclose each r_i in an interval (a_i, b_i) whose length is $\frac{\epsilon}{2^i}$. Thus Q is contained in intervals whose total length is no greater than $\frac{\epsilon}{2^0} + \frac{\epsilon}{2^1} + \frac{\epsilon}{2^2} + \cdots \frac{\epsilon}{2^i} + \cdots$. From the Appendix, the sum of this series is 2ϵ. But since ϵ can be any positive number, however small, we see that the "measure" of Q can be nothing other than zero.

Now, if $m(Q) = 0$, then $m(I) = 1$, since $m([0, 1]) = 1$. So Lebesgue's sum S is $0 \cdot 0 + 1 \cdot 1 = 1$. Thus we should conclude that the Lebesgue integral of Dirichlet's function is 1. Notice that according to Lebesgue's theory, the values of f at rational numbers have no effect on the "area" under Dirichlet's f!

Exercise 3.32: Prove that $\lim \frac{\sin \alpha}{\alpha} = 1$, as α approaches zero, as Cauchy does in his Lesson One.

Exercise 3.33: Prove that $(1 + \alpha)^{\frac{1}{\alpha}}$ is a number between 2 and 3, which Cauchy does in his Lesson One, and says he has calculated to be 2.7183 to within an accuracy of one ten-thousandth by taking $\alpha = 1/10,000$ and using a table of decimal logarithms. This limit, called e, is as important as π in mathematics. Hint: First restrict to $\alpha = 1/m$ for m a natural number, and expand via the Binomial Theorem. Try to give an argument for why this must have a limit, and why it lies between 2 and 3. Consult Cauchy's original text as needed.

Exercise 3.34: Cauchy studies whether the functions x^m, $1/x^m$, A^x, $L(x)$, $\sin x$, $\cos x$ are continuous (here m is a natural number, A is a positive number, and $L(x)$ is the logarithm function with base A). Show that these functions are continuous, discussing any difficulties you (or Cauchy) might encounter.

Exercise 3.35: Follow Cauchy by calculating the derivatives of the functions $a+x$, $a - x$, ax, a/x, $\sin x$, $\cos x$, where a is any constant real number.

Exercise 3.36: Using the notation in Cauchy's definition and Exercise 3.34, follow him by calculating the derivatives of

    (a) $y = L(x)$ (Hint: Let $i = \alpha x$).
    (b) $y = A^x$ (Hint: Let $A^i = 1 + \beta$).
    (c) $y = x^a$ (Hint: Let $i = \alpha x$, $(1 + \alpha)^a = 1 + \gamma$).

What restrictions do you need to place on x and a?

When A is chosen to be e, Cauchy denotes $L(x)$ by $l(x)$, which he calls the *Napierian* or *hyperbolic* logarithm (Find out why!). Today we call it the *natural* logarithm. What is the form of the derivatives in parts (a) and (b) in this case?

Exercise 3.37: Cauchy develops a technique for calculating the derivative of a composite function, i.e., a function of a function. Thus if $z = F(y)$ and $y = f(x)$, Cauchy deduces (thinking of z as the composite function of x) $z' = f'(x) \cdot F'(f(x))$. Produce a simple proof of this, as Cauchy did, but discuss how the proof may fail in certain situations.

Today we call this formula the *chain rule*, since one can use it to differentiate any function formed as a composite chain of simpler functions. Follow Cauchy by computing the derivative of $\tan x$, and then use the chain rule to find formulas for the derivatives of the inverse trigonometric functions arcsin, arccos, arctan, noting any restrictions necessary.

Exercise 3.38: Find formulas for the derivatives of functions formed as sums, differences, products, or quotients of other functions.

Exercise 3.39: Learn enough about computing derivatives (e.g., from the preceding exercises, or by reading Cauchy) to find all the successive derivatives at $x = 0$ of the function defined to equal zero at zero and e^{-1/x^2} elsewhere. Can you explain how this makes the point that two different functions can have the same power series representation?

Exercise 3.40: For $f(x)$ always positive, explain how Cauchy's sum S in Lesson 21 represents an approximation for the area under a curve.

Exercise 3.41: (a) From Cauchy's definition of the definite integral, calculate $\int_{x_0}^{X} a \, dx$ and $\int_{x_0}^{X} x \, dx$ (Hint: $1 + 2 + 3 + \cdots + n = n(n+1)/2$), and $\int_{x_0}^{X} A^x \, dx$, by choosing subintervals of equal width.

(b) By choosing subintervals with equal ratio between successive subintervals, calculate $\int_{x_0}^{X} x^a \, dx$. (Hint: Let the ratio be $1 + \alpha$ with α small and positive. Also review Exercise 3.36c.)

Exercise 3.42: Calculate $\int_0^1 dx / \left(1 + x^2\right)$ to within an accuracy of $1/16$.

Exercise 3.43: Verify, with Cauchy, that $\int_0^1 dx / \left(1 + x^2\right) = \pi/4$. (Hint: In Exercise 3.37 you computed $(\arctan x)'$.)

Exercise 3.44: Use Cauchy's Fundamental Theorem of Calculus to recover the Quadrature of the Parabola, by Archimedes, and the area under a segment of $y = x^n$, as calculated earlier by Leibniz's approach.

3.7   Robinson Resurrects Infinitesimals

Our final source displays a surprising leap from the seventeenth to the twentieth century, in which Abraham Robinson has resurrected the infinitesimals of Leibniz that Cauchy's nineteenth-century successors had worked so assiduously to eliminate from analysis.

Robinson was born in 1918 in the Silesian town Wałbrzych (Waldenburg, Prussia), which is today in southwestern Poland [111]. (The recent biography of Robinson by J. Dauben is fascinating reading [37].) As the Nazi dictatorship rose in Germany, and Jewish families like his were suffering escalating persecution, his family emigrated to Palestine, where Robinson studied with Abraham Fraenkel, whose work on the foundations of mathematics is discussed in the set theory chapter.

Robinson came to the Sorbonne, in Paris, in 1939 to study further, but evacuated to England when the Nazis invaded France. He studied at the University of London, but soon joined de Gaulle's French Free Air Force, and then the British Air Force, where he did theoretical work on airplane design. After the war he taught mathematics at Cranfield College of Aeronautics, and completed his Ph.D. at the University of London with a dissertation on "The Meta-mathematics of Algebraic Systems." He became a professor of applied mathematics at the University

PHOTO 3.7. Robinson.

of Toronto, then professor of mathematics at the Hebrew University in Jerusalem; University of California, Los Angeles; and finally Yale University. Robinson first made substantial contributions to aerodynamics, particularly to subsonic and supersonic airplane wing behavior, and the spread of pressure waves. Later he became famous for his work on mathematical model theory in mathematical logic, which included his discoveries about infinitesimals. It was in Los Angeles that he finished his book on the subject, *Non-Standard Analysis* [140, Preface and pages 2, 282].

Abraham Robinson, from
Non-Standard Analysis

In the fall of 1960 it occurred to me that the concepts and methods of contemporary Mathematical Logic are capable of providing a suitable framework for the development of the Differential and Integral Calculus by means of infinitely small and infinitely large numbers....

[T]he idea of infinitely small or infinitesimal quantities seems to appeal naturally to our intuition. At any rate, the use of infinitesimals was widespread during the formative stages of the Differential and Integral Calculus. As for the objection...that the distance between two distinct real numbers cannot be infinitely small, G.W. Leibniz argued that the theory of infinitesimals implies the introduction of ideal numbers which might be infinitely small or infinitely large compared with the real numbers, but which were to possess the same properties as the latter. However, neither he nor his disciples and successors were able to give a rational development leading up to a system of this sort. As a result, the theory of infinitesimals gradually fell into disrepute and was replaced eventually by the classical theory of limits.

It is shown in this book that Leibniz's ideas can be fully vindicated and that they lead to a novel and fruitful approach to classical Analysis and to many other branches of mathematics. The key to our method is provided by the detailed analysis of the relation between mathematical languages and mathematical structures which lies at the bottom of contemporary model theory....

[T]he infinitely small and infinitely large numbers of a non-standard model of Analysis are neither more nor less real than, for example, the standard irrational numbers.

What Robinson created in his book was a completely rigorous new system of numbers containing not only the "standard" real numbers, properly defined in the late nineteenth century, but also a plethora of additional "infinitesimal" numbers (and their infinitely large reciprocals as well), which Leibniz had supposed and used without justification. Together these are called the "nonstandard real numbers."

Robinson believed that his nonstandard analysis would not only rewrite calculus as a mathematical subject, but that it should rewrite the history of calculus too [37, p. 355]. He presented his view of this revision in Chapter X of his book. Robinson was interested "in getting into Leibniz's mind" [37, p. 349], and spoke of Newton and Leibniz as if he knew them [37, p. 349]. He wrote [37, p. 350]:

One may ask, if Leibniz' ideas are indeed tenable, how is it that they degenerated in the eighteenth century and suffered eclipse in the nineteenth? The reason for this can hardly be found in the absence of a consistency proof for his system. Leibniz was aware of a need for justifying his procedure and he did so by relying on the fact that "tout se gouverne par raison" [everything is controlled by ratio]— hardly a consistency proof in our sense. However, even in the twentieth century no mathematician is known to have changed his profession because Gödel[10] showed that no conclusive proof for the consistency of arithmetic is possible. Nor can it be said that a theory of infinitesimals is less intuitive than the δ, ϵ-procedure[11] and its ramifications and developments.

The reason for the eventual failure of the theory is to be found rather in the fact that neither Leibniz nor his successors were able to state with sufficient precision just what rules were supposed to govern their system of infinitely small and infinitely large numbers.

Robinson's biographer Joseph Dauben continues:

There is a major difference, however, between Gödel's result and Leibniz's imprecision regarding the infinitesimal calculus, for Gödel's work never cast any doubts on the validity of arithmetic per se—although it did dash hopes

---

[10]Kurt Gödel was the predominant mathematical logician of the early twentieth century.

[11]This is the method developed by Weierstrass and others for doing calculus without infinitesimals.

that the consistency of arithmetic could ever be *proven*. On the other hand, the dual problems of the infinite and infinitesimals had always plagued mathematicians and philosophers alike because they always seemed to be involved with inconsistencies and paradoxes. While there were no good grounds to doubt the consistency of arithmetic, there were historically long-standing reasons to be suspicious of infinitesimals, for reasons as well known to the Greeks as they were to...Berkeley.... It was the nagging worry that contradictions could not be explained away that made their case very different from Gödel's [37, pp. 350–51].

On the validity, acceptance, use, and importance of number systems in general, Robinson quipped, in an informal article "Numbers—What Are They and What Are They Good For?" [37, pp. 472–73][141], that

Number systems, like hair styles, go in and out of fashion—it's what's underneath that counts.

and, in a more serious vein, that he wanted to demonstrate that

The collection of all number systems is not a finished totality whose discovery was complete around 1600, or 1700, or 1800, but that it has been and still is a growing and changing area, sometimes absorbing new systems and sometimes discarding old ones, or relegating them to the attic.

Robinson, one of the great mathematicians of the twentieth century, died in 1974 in New Haven, Connecticut.

And thus we are brought more than full circle in our story of one of the major controversies in the philosophy of mathematics: first the Greek method of exhaustion rejecting the infinitely small or large, to the mysterious but attractive infinitesimals of Cavalieri, Leibniz, and Newton, then their complete banishment by the nineteenth century rigorization of Cauchy, Weierstrass, and others, and now finally their resurrection by Robinson. Who knows what twists and turns the future will bring in this and other parts of mathematics?

One view of how Robinson's discovery may connect the past and future comes from Kurt Gödel himself:

I would like to point out a fact that was not explicitly mentioned by Professor Robinson, but seems quite important to me; namely that non-standard analysis frequently simplifies substantially the proofs, not only of elementary theorems, but also of deep results.... [T]here are good reasons to believe that non-standard analysis, in some version or other, will be the analysis of the future.

One reason is the just mentioned simplification of proofs, since simplification facilitates discovery. Another, even more convincing reason, is the following: Arithmetic starts with the integers and proceeds by successively enlarging the number system by rational and negative numbers, irrational numbers, etc. But the next quite natural step after the reals, namely the introduction of infinitesimals, has simply been omitted. I think, in coming centuries it will be considered a great oddity in the history of mathematics that the first exact theory of infinitesimals was developed 300 years after the invention of the differential calculus. I am inclined to believe that this oddity has something to do with another oddity relating to this span of time, namely the fact that such problems as Fermat's,[12] which can be written down in ten symbols of elementary arithmetic, are still unsolved 300 years after they have been posed. Perhaps the omission mentioned is largely responsible for the fact that, compared to the enormous development of abstract mathematics, the solution of concrete numerical problems was left far behind [140, Preface to the second edition].

3.8   Appendix on Infinite Series

We give a very brief, informal introduction to infinite sums, which are today called infinite series. Any calculus book will take the reader further.

At first sight it seems that an infinite sum, such as

$$1 + \frac{1}{2} + \frac{1}{4} + \frac{1}{8} + \cdots + \frac{1}{2^i} + \cdots,$$

could have no meaning, first because it involves completing infinitely many additions all at once, second because it seems that the final result could not possibly be finite.

The first problem is resolved by considering only finitely many terms at once, by letting $s_i = 1 + \frac{1}{2} + \frac{1}{4} + \frac{1}{8} + \cdots + \frac{1}{2^i}$, and calling s_i the ith partial sum. The final result of the entire summation is then interpreted as the limiting behavior of the sequence of numbers s_i as i "approaches infinity"; i.e., we confront the second problem by asking whether the numbers s_i approach a fixed limiting value as i grows indefinitely. In our example, we can compute $s_i = 1 + \frac{1}{2} + \frac{1}{4} + \frac{1}{8} + \cdots + \frac{1}{2^i} = 2 - \frac{1}{2^i}$ (Check this algebraically!). Thus it seems clear that the partial sums $s_i = 2 - \frac{1}{2^i}$ actually do approach a fixed number, namely 2, as i grows indefinitely, so we write $1 + \frac{1}{2} + \frac{1}{4} + \frac{1}{8} + \cdots + \frac{1}{2^i} + \cdots = 2$, and say that the infinite series converges and has sum 2 (Exercise 3.45). Whether or not a particular infinite series actually converges is very delicate, and often not easy to determine (Exercise 3.46).

We can use infinite series to represent functions by introducing a variable x, as in the sample "power series" function

$$f(x) = 1 + \frac{x}{2} + \frac{x^2}{4} + \frac{x^3}{8} + \cdots + \frac{x^i}{2^i} + \cdots.$$

In fact, this series, looking like an "infinite polynomial" involving only coefficients times powers of x, represents the function $\frac{1}{1-\frac{x}{2}}$ when $|x| < 2$, and fails to converge otherwise (Exercise 3.47). The form of a power series representation of a function is intimately tied to calculus, e.g., the derivative and antiderivative (Exercise 3.48).

---

[12]Fermat's Last Theorem; see the number theory chapter.

Exercise 3.45: Calculate: $1 + \frac{1}{3} + \frac{1}{9} + \frac{1}{27} + \cdots + \frac{1}{3^i} + \cdots$

Exercise 3.46: Show that while the infinite series $1 + \frac{1}{2} + \frac{1}{4} + \frac{1}{8} + \cdots$ can reasonably be said to "converge to," or represent, the number 2, the series $1 + \frac{1}{2} + \frac{1}{3} + \frac{1}{4} + \cdots$ fails to converge to any number, i.e., it "diverges." (Hint: $\frac{1}{3} + \frac{1}{4} > \frac{1}{4} + \frac{1}{4}$ and $\frac{1}{5} + \frac{1}{6} + \frac{1}{7} + \frac{1}{8} > \frac{1}{8} + \frac{1}{8} + \frac{1}{8} + \frac{1}{8}$.) This particular divergent series is so important that it has a special name, the "harmonic series." Its divergence had been demonstrated already by the French cleric and mathematician Nicole Oresme in the fourteenth century!

Exercise 3.47: Show that

$$1 + \frac{x}{2} + \frac{x^2}{4} + \frac{x^3}{8} + \cdots + \frac{x^i}{2^i} + \cdots = \frac{1}{1 - \frac{x}{2}}$$

when $|x| < 2$, but that the infinite series fails to converge otherwise.

Exercise 3.48: From Exercises 3.35 and 3.36 of the section on Cauchy, we know the derivatives of four important functions: $\sin'(x) = \cos(x)$, $\cos'(x) = -\sin'(x)$, $(e^x)' = e^x$ and $l'(x) = \frac{1}{x}$. And from Cauchy's Third Lesson we know that $(x^m)' = mx^{m-1}$. Assuming that a power series can be differentiated term by term, i.e., that the derivative of the sum is the sum of the derivatives, find power series representations for these four important functions.

CHAPTER 4

Number Theory: Fermat's Last Theorem

4.1 Introduction

On June 24, 1993, the *New York Times* ran a front-page story with the headline "At Last, Shout of 'Eureka!' In Age-Old Math Mystery." The proverbial shout of "Eureka!" had echoed across the campus of Cambridge University, England, just the day before. At the end of a series of lectures at a small conference on the arcane subjects of "*p*-adic Galois Representations, Iwasawa Theory, and the Tamagawa Numbers of Motives," Princeton mathematician Andrew Wiles mentioned, almost as an afterthought, that the results he had presented implied, as a corollary, that Fermat's Last Theorem was true. Via telephone and electronic mail, the news of what many mathematicians called the most exciting event in twentieth-century mathematics spread around the globe almost instantly. We will return to these developments again at the end of this introductory section.

Fermat's Last Theorem (FLT), the focus of all this commotion, is easily stated, saying that the equation $x^n + y^n = z^n$ has no solution in terms of nonzero integers x, y, z, if the integer exponent n is greater than two. Until that day in June 1993 this statement might more appropriately have been called a conjecture, since it had remained unproven, despite the efforts of some of the world's best mathematicians for three hundred years since Pierre de Fermat's bold claim during the first half of the seventeenth century. Their efforts helped develop an entire new branch of mathematics.

Who was Fermat and what led him to make such a curious assertion? The Frenchman Pierre de Fermat (1601–1665) was one of the truly great figures in the history of mathematics. With his work he made essential contributions to the transition from the classical Greek tradition to a wholly new approach to mathematics, which took place in Europe during the seventeenth century. Much of the sixteenth and early seventeenth century was devoted to translating into Latin, restoring, and

PHOTO 4.1. Wiles beside the Fermat memorial in Beaumont-de-Lomagne, Fermat's birthplace.

extending mathematics texts from classical Greece, such as the works of Euclid, Apollonius, Archimedes, Pappus, Ptolemy, and Diophantus of Alexandria. Fermat himself undertook several such restoration projects, such as Apollonius's *Plane Loci*. Even in the early seventeenth century, they were viewed as the pinnacle of mathematical achievement.

The mathematical community during the seventeenth century was quite different from what it is today. There was nothing like a mathematical profession, with professional standards and established methods of publication and communication. What is more, mathematics did not even have a clear identity as a separate discipline, and there was no agreement as to what it should be. Hardly anybody was making a living doing mathematical research, and scholars pursued mathematics for a variety of different reasons. The only mathematics taught at universities was some basics necessary for degrees in law, medicine, or theology. Descriptions of Fermat's life usually emphasize that he was an "amateur," which makes his accomplishments seem all the more astounding. But obviously the term cannot really be meaningfully applied to the time period he lived in.

PHOTO 4.2. Fermat.

Fermat received a law degree from the University of Orléans, France, in 1631, after which he moved to Toulouse, where he lived the rest of his life, traveling regularly to other cities. He practiced law and soon became a "councillor" to the "Parlement," the provincial High Court in Toulouse, a position he kept until his death. Thus, his mathematical research was done in his spare time, and there were long periods during his life when his professional duties kept him from seriously pursuing research. There are many indications that Fermat did mathematics partly as a diversion from his professional duties, for personal gratification. While he enjoyed the attention and esteem he received from many of his mathematical peers, he never showed interest in publishing his results. He never traveled to the centers of mathematical activity, not even Paris, preferring to communicate with the scientific community through an exchange of letters, facilitated by the theologian Marin Mersenne (1588–1648), in Paris, who served as a clearing house for scientific correspondence from all over Europe.

The central mathematical influence in Fermat's life was François Viète (1540–1603) and his school in Bordeaux. He became acquainted with disciples of Viète during a long stay in Bordeaux in the late 1620s. In 1591, Viète had published his *Introduction to the Analytic Art*, the first in a series of treatises, in which he outlined a new system of symbolic algebra, promising a novel method of mathematical discovery. As he says in the introduction:

There is a certain way of searching for the truth in mathematics that Plato is said first to have discovered. Theon called it analysis, which he defined as assuming that which is sought as if it were admitted [and working] through the consequences [of that assumption] to what is admittedly true, as opposed to synthesis, which is assuming what is [already] admitted [and working] through the consequences [of that assumption] to arrive at and to understand that which is sought.

Although the ancients propounded only [two kinds of] analysis, zetetics and poristics, to which the definition of Theon best applies, I have added a third, which may be called rhetics or exegetics. It is properly zetetics by which one sets up an equation or proportion between a term that is to be found and the given terms, poristics by which the truth of a stated theorem is tested by means of an equation or proportion, and exegetics by which the value of the unknown term in a given equation or proportion is determined. Therefore the whole analytic art, assuming this three-fold function for itself, may be called the science of correct discovery in mathematics [172, pp. 11–12].

Viète's work represents an important milestone in the transition from ancient to modern mathematics, even though he was not a great influence on the scientific community at the time, and his symbolic algebra was soon eclipsed by the work of René Descartes (1596–1650). (More details about Viète's work can be found in [42] and [93]. The influence of Viète on Fermat is described in detail in [113, Ch. II].) Fermat adopted Viète's symbolic algebra and adhered to it in all his writings. Viète's theory of equations formed the launching pad for Fermat's work in number theory and analysis.

While Fermat made very important contributions to the development of the differential and integral calculus (see the analysis chapter and [113, Ch. IV]), and to analytic geometry [113, Ch. III], his lifelong passion belonged to the study of properties of the integers, now known as number theory, and it is there that Fermat had the most lasting influence on the course of mathematics in later centuries. His number-theoretic research is centered on just a handful of themes, rooted in the classical Greek tradition, involving the notions of divisibility and primality.

First, Fermat focused on the problem of finding *perfect numbers*, those numbers that are equal to the sum of their proper divisors. For instance, $6 = 1 + 2 + 3$ is perfect. This problem had already occupied the Pythagoreans, and the main classical Greek achievement is recorded as Proposition 36 in Book IX of Euclid's *Elements*: *If as many numbers as we please beginning from a unit be set out continuously in double proportion, until the sum of all becomes prime, and if the sum multiplied into the last make some number, the product will be perfect.* In modern terms, this proposition asserts that, if $2^{n+1} - 1$ is prime for some integer $n \geq 1$, then $2^n(2^{n+1} - 1)$ is a perfect number (Exercise 4.1). (Why is this statement equivalent to Proposition 36?) But the problem is far from solved, because it remains open whether there are other perfect numbers not of this form. More importantly, however, to use the proposition to find perfect numbers, one needs an efficient way to test whether a given number is prime. While Fermat made substantial progress on the latter, it was not until the eighteenth century that Euler proved that all even perfect numbers are of the form given in Euclid's proposition. The question whether there are any odd perfect numbers remains one of the important unsolved problems in number theory today. It is known that there is no odd perfect number less than 10^{160} [71, p. 167]. For a historical survey of work on perfect numbers see [41, vol. I, Ch. 1], [131, Ch. 5].

In any case, by Euclid's proposition, every prime number in the sequence

$$2^2 - 1, 2^3 - 1, 2^4 - 1, \ldots, 2^n - 1, \ldots$$

produces a perfect number. Such primes are now known as *Mersenne primes*. Fermat's major tool to test primality of these numbers is now known as Fermat's Theorem (sometimes called Fermat's Little Theorem). It says, in his own words:

> Without exception, every prime number measures one of the powers -1 of any progression whatever, and the exponent of the said power is a submultiple of the given prime number -1. Also, after one has found the first power that satisfies the problem, all those of which the exponents are multiples of the exponent of the first will similarly satisfy the problem [113, p. 295].

This theorem is stated today (Exercise 4.2) as follows:

Fermat's Theorem: Given a prime number p and an integer a that is not divisible by p, then a^{p-1} has remainder 1 under division by p. Furthermore, there exists a least positive integer n such that a^n has remainder 1 under division by p, n divides $p - 1$, and a^{kn} has remainder 1 under division by p for all positive integers k.

How does this result help? First of all, observe that if $2^n - 1$ is prime, then n itself has to be prime (Exercise 4.3). Fermat then drew the following corollary from his theorem (Exercise 4.4), which greatly limits the number of potential divisors of $2^n - 1$ to be checked.

Corollary: Let p be an odd prime, and q a prime. If q divides $2^p - 1$, then q is of the form $2kp + 1$ for some integer k.

For large primes p, this method will still be rather slow, and quicker methods have since been developed [71, p. 171].

A very surprising application of Fermat's Little Theorem surfaced in the 1970s, when it was applied to the construction of very secure secret codes, so-called public key cryptosystems. These have found ubiquitous uses in information transfer in business and banking, including automatic teller machines. For this and other applications of number theory see [153].

Fermat then broadened his investigation of primality to numbers of the form $a^n + 1$, for integers a and n. A letter to Mersenne, dated Christmas Day 1640, suggests that he found a proof that such a number could be prime only if a is even and n is a power of 2 (Exercise 4.5). Based on his calculations, Fermat conjectured that in fact all numbers of the form $2^{2^n} + 1$ are prime. A proof of this conjecture seemed to elude him for many years, until he wrote in a letter to his correspondent Carcavi in 1659 that he had finally found it [113, p. 301]. In 1732, Leonhard Euler showed that $2^{2^5} + 1$ is divisible by 641. Primes of this form are now known as *Fermat primes*.

Besides the study of perfect numbers, the other important source of inspiration for Fermat's number-theoretic researches was the *Arithmetica* of Diophantus of

Alexandria, who lived during the third century. He was one of the last great mathematicians of Greek antiquity. The *Arithmetica* is a collection of 189 problems relating to the solution of equations in one or more variables taken to be fractions, originally divided into thirteen books, of which only six are preserved [9].[1] The solutions are presented in terms of specific numerical examples, with rational numbers. An instance of relevance to the present chapter is Problem 8 from Book II, taken from [93, p. 166] (in modernized notation):

Problem II-8. *To divide a given square number into two squares.*
Let it be required to divide 16 into two squares. And let the first square $= x^2$; then the other will be $16 - x^2$; it shall be required therefore to make $16 - x^2 =$ a square. I take a square of the form $(ax - 4)^2$, a being any integer and 4 the root of 16; for example, let the side be $2x - 4$, and the square itself $4x^2 + 16 - 16x$. Then $4x^2 + 16 - 16x = 16 - x^2$. Add to both sides the negative terms and take like from like. Then $5x^2 = 16x$, and $x = 16/5$. One number will therefore be $256/25$, the other $144/25$, and their sum is $400/25$ or 16, and each is a square.

Clearing denominators, one easily obtains an integer solution to this type of problem. In this vein, triples of integers x, y, z that satisfy the equation

$$x^2 + y^2 = z^2$$

are called *Pythagorean triples* (Exercise 4.6). Examples are $(3, 4, 5)$ and $(5, 12, 13)$. The search for Pythagorean triples goes back at least to the Babylonians. Our first source in this chapter comes from Euclid's *Elements*, in which he gives a complete description of all (infinitely many) Pythagorean triples. Via the Pythagorean Theorem, such triples correspond, of course, to right triangles with integer sides (Exercises 4.7, 4.8, 4.9).

Diophantus obtains challenging variations of this problem by requiring solutions that satisfy extra conditions, such as Problem 6 in Book VI, which asks for a right triangle (with rational sides) such that the sum of its area and one of the legs of the right angle is equal to a given number [113, pp. 304 f.]. (See also [40, vol. II, pp. 176 ff.].) Fermat greatly extended Diophantus's method of "single and double equations," as it was called, and made it into a powerful weapon to solve most problems of this type.

Another line of research Fermat pursued, which was destined to be investigated in great depth by later generations of number theorists, again starts with a problem from Diophantus's *Arithmetica*. Problem 19 in Book III asks for four numbers such that if any one of them is added to, or subtracted from, the square of their sum, the result is a square. Diophantus reduces the problem to finding four right triangles with a common hypotenuse. He then proceeds to give a specific numerical solution [113, p. 315]. Now, the edition of the *Arithmetica* that Fermat was basing his research on had been published by Claude Gaspar Bachet de Méziriac in 1621. Bachet had developed an interest in mathematical recreations and puzzles and,

---

[1] In 1972, R. Rashed found four more books of the *Arithmetica* with 101 additional problems in the library of the tomb of the Imam Resa in Mashad, Iran. see [136, 155].

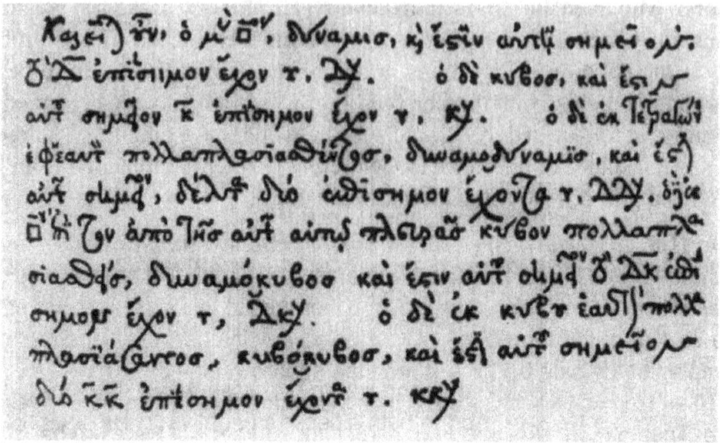

PHOTO 4.3. Diophantus on equations, from a fourteenth-century manuscript.

drawn to number theory, prepared a new translation from Greek into Latin, the scientific lingua franca of the era, along with an annotation of the *Arithmetica*. Bachet, in pursuit of a general solution to Problem 19, reduced the question further to that of how to find numbers that were sums of two squares in a prescribed number of ways. In his commentary, he gives some specific answers but no general solution. Once again, Fermat's genius brings forth a complete solution to the problem. He uses the now common approach of reducing the problem to considering prime numbers first, and building up the general solution via the factorization of a given number into its prime factors. Odd prime numbers can be divided into two classes, those of the form $4k - 1$, for some integer $k \geq 1$, and those of the form $4k + 1$. He shows that primes of the former kind cannot be the sum of two squares, and play no role in the general solution. The solution is given as follows, in Fermat's own words.

A prime number, which exceeds a multiple of four by unity, is only once [i.e., in one way] the hypotenuse of a right triangle, its square twice, its cube three times, its quadratoquadrate [fourth power] four times, and so on *in infinitum*.... [113, p. 316].

He then investigates products of primes of the form $4n + 1$, and without indication of his method of proof, Fermat then makes the (correct) claim that, if $n = n' p_1^{a_1} p_2^{a_2} \cdots p_r^{a_r}$, where the p_i are primes of the form $4k + 1$, and n' is composed of prime factors of the form $4k - 1$, then n^2 can be written as a sum of two squares in

$$\tfrac{1}{2}[(2a_1 + 1)(2a_2 + 1) \cdots (2a_r + 1) - 1]$$

ways [113, pp. 318 f.]. (See also [177, p. 71].) Today results of this kind form part of what we call the theory of quadratic forms. An excellent book on sums of squares is [80].

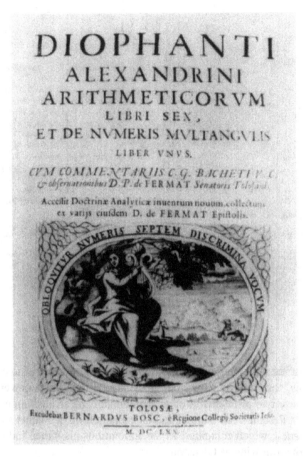

PHOTO 4.4. Frontispiece from Samuel Fermat's edition.

At the time, Fermat did not reveal the proof of this result. Only some years later, in 1659, in a letter to Christian Huygens (1629–1695), the inventor of the pendulum clock, did he finally reveal the method he had used to prove this and many other spectacular results, which he called the "method of infinite descent." He illustrates it for Huygens by outlining a proof that there is no right triangle whose area is a square integer. If there were such a triangle, then he could construct another right triangle whose area is square, but smaller than the area of the first triangle. In turn, he could begin with the newly constructed triangle and find yet another one with smaller area a square, and so on. But since this process, which results in smaller and smaller positive integers, cannot go on forever, one could not have been able to find the first triangle that got it started. While this method would seem to be suitable only for proving negative results, that certain things are impossible, Fermat was able to adapt it to prove positive statements, such as the above assertion that every prime of the form $4k + 1$ is a sum of squares.

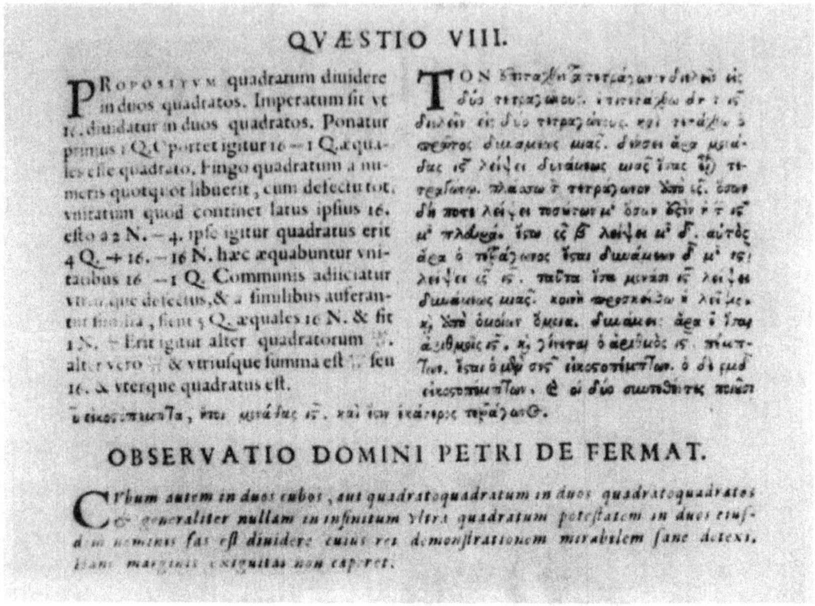

QVÆSTIO VIII.

PROPOSITVM quadratum diuidere in duos quadratos. Imperatum sit vt 16. diuidatur in duos quadratos. Ponatur primus 1 Q. Oportet igitur 16 — 1 Q. æquales esse quadrato. Fingo quadratum a numeris quotquot libuerit, cum defectu tot vnitatum quod continet latus ipsius 16. esto a 2 N. — 4. ipse igitur quadratus erit 4 Q. + 16. — 16 N. hæc æquabuntur vnitatibus 16 — 1 Q. Communis adiiciatur vtrinque defectus, & a similibus auferantur similia, fient 5 Q. æquales 16 N. & fit 1 N. Erit igitur alter quadratorum 16/25. alter vero 144/25 & vtriusque summa est 400/25 seu 16. & vterque quadratus est.

ΤΟΝ ἐπιταχθέντα τετράγωνον διελεῖν εἰς δύο τετραγώνους. ...

OBSERVATIO DOMINI PETRI DE FERMAT.

CVbum autem in duos cubos, aut quadratoquadratum in duos quadratoquadratos & generaliter nullam in infinitum vltra quadratum potestatem in duos eiusdem nominis fas est diuidere cuius rei demonstrationem mirabilem sane detexi. Hanc marginis exiguitas non caperet.

PHOTO 4.5. Fermat's marginal comment.

It seems that the same method allowed him to prove that there is no cube that is a sum of cubes, nor a fourth power that is the sum of two fourth powers. Earlier, he had sent these two problems to other mathematicians as challenge problems. When, in 1670, Fermat's son Samuel published an edition of Bachet's translation of the *Arithmetica*, which contained all the annotations his father had made in it, one can find the following as Observation 2:

No cube can be split into two cubes, nor any biquadrate into two biquadrates, nor generally any power beyond the second into two of the same kind [177, p. 104].

In other words, Fermat claims that the equation $x^n + y^n = z^n$ has no nonzero integer solutions when n is greater than 2. Tantalizingly, he added that the margin was too narrow to contain the truly remarkable proof, an explanation used by him also elsewhere to explain the absence of a proof. This most famous marginal note has become known as "Fermat's Last Theorem" and has occupied mathematicians ever since, culminating in the proof by Andrew Wiles. Naturally, the question whether Fermat indeed had a proof or just naively assumed that his method of infinite descent would generalize for all exponents has been much discussed. Following are the opinions of two of the leading mathematicians of the twentieth century. First, André Weil remarks:

As we have observed...the most significant problems in Diophantus are concerned with curves of genus 0 or 1. With Fermat this turns into an almost exclusive concentration on such curves. Only on one ill-fated occasion did Fermat ever mention a curve of higher genus, and there can hardly remain any doubt that this was due to some misapprehension on his part, even though, by a curious twist of fate, his reputation in the eyes of the ignorant came to rest chiefly upon it. By this we refer of course to the incautious words "*et generaliter nullam in infinitum potestatem*" in his statement of "Fermat's last theorem" as it came to be vulgarly called.... How could he have guessed that he was writing for eternity? We know his proof for biquadrates...he may well have constructed a proof for cubes, similar to the one which Euler discovered in 1753...he frequently repeated those two statements...but never the more general one. For a brief moment perhaps, and perhaps in his younger days...he must have deluded himself into thinking that he had the principle of a general proof; what he had in mind on that day can never be known [177, p. 104].

A more cautious opinion was expressed by L.J. Mordell [126, p. 4]:

Mathematical study and research are very suggestive of mountaineering. Whymper made seven efforts before he climbed the Matterhorn in the 1860s and even then it cost the lives of four of his party. Now, however, any tourist can be hauled up for a small cost, and perhaps does not appreciate the difficulty of the original ascent. So in mathematics, it may be found hard to realise the great initial difficulty of making a little step which now seems so natural and obvious, and it may not be surprising if such a step has been found and lost again.

In hindsight, Fermat was one of the great mathematical pioneers, who built a whole new paradigm for number theory on the accomplishments of classical Greece, and laid the foundations for a mathematical theory that would later be referred to as the "queen of mathematics." But, as is the fate of many scientific pioneers, during his lifetime he tried in vain to interest the scientific community in his number-theoretic researches. After unsuccessful attempts to interest such leading mathematicians as John Wallis (1616–1703), a very influential predecessor of Newton in England, and Blaise Pascal (1623–1662) in Paris, Fermat made a last attempt to win over Huygens, in the letter referred to above. He concludes the letter as follows:

There in summary is an account of my thoughts on the subject of numbers. I wrote it only because I fear I shall lack the leisure to extend and to set down in detail all these demonstrations and methods. In any case, this indication will serve learned men in finding for themselves what I have not extended, particularly if MM. de Carcavi and Frénicle share with them some proofs by infinite descent that I sent them on the subject of several negative propositions. And perhaps posterity will thank me for having shown it that the ancients did not know everything, and this relation will pass into the mind of those who come after me as a "passing of the torch to the next generation," as the great Chancellor of England says, following the sentiment and the device of whom I will add, "Many will pass by and knowledge will increase" [113, p. 351].

Whether it was the sentiment of the times, or Fermat's secretiveness about his methods of discovery and proofs of results that he presented only as challenges, he was singularly unsuccessful in enticing the great minds among his contemporaries to follow his path. It was to be a hundred years before another mathematician of Fermat's stature took the bait and carried on Fermat's work.

Leonhard Euler (1707–1783) was without doubt one of the greatest mathematicians the world has ever known. A native of Switzerland, Euler spent his working life at the Academies of Sciences in St. Petersburg and Berlin. His mathematical interests were wide-ranging, and included number theory, which he is said to have pursued as a diversion, in contrast to the more mainstream areas of research to which he contributed. It was Christian Goldbach (1690–1764) who drew Euler's attention to the works of Fermat, beginning with their very first exchange, in 1729, initiated by Euler. In his reply, Goldbach adds as a postscript: "Is Fermat's observation known to you, that all numbers $2^{2^n} + 1$ are primes? He said he could not prove it; nor has anyone else done so to my knowledge" [177, p. 172]. Their correspondence was to last more than thirty years, until Goldbach's death. Goldbach was a well-traveled and well-educated man whose main intellectual interests were languages and mathematics. He knew many of the distinguished mathematicians of his time, including Nicolas (1687–1759) and Daniel Bernoulli (1700–1782), both of whom obtained appointments to the Academy in St. Petersburg, thanks to his efforts. They, in turn, managed to get an appointment there for the young Euler.

A large part of Euler's number-theoretic work consisted essentially in a systematic program to provide proofs for all the assertions of Fermat [177, p. 170], including Fermat's Last Theorem (FLT). He provided the first proof for exponent three, which is considerably harder than that for four. The second original source in this chapter is Euler's own proof for exponent four.

There was only a small number of scholars during the second half of the eighteenth century who were interested in pure mathematics. Fortunately, one of them devoted part of his career to the pursuit of number theory. In 1768, Joseph Louis Lagrange (1736–1813) became interested in number theory and produced a string of publications on this subject during the following decade; much of it was directly inspired by Euler's work. No new results on FLT emerged from his publications, however, but he carried on the number-theoretic tradition, to be taken up by later researchers. Lagrange had become the successor of Euler at the Academy of Sciences in Berlin, and inherited the role of foremost mathematician in Europe after Euler's death. (For a biographical sketch of Lagrange see the algebra chapter.) In 1786, Lagrange left Berlin for Paris, where he was to spend the rest of his life.

One of his colleagues there was Adrien-Marie Legendre (1752–1833), who had attracted Lagrange's attention four years earlier, when Legendre sent him a prizewinning essay on ballistics [177, p. 324]. (See the geometry chapter for more information on Legendre.) In 1785, Legendre submitted to the Paris Academy an essay entitled *Researches on Indeterminate Analysis*, containing his first work on number theory, directly inspired by the writings of Euler and Lagrange. By that time, Euler was dead and Lagrange was no longer actively working in this area. Legendre embarked on an extensive number-theoretic research program, which resulted in a comprehensive treatment of number theory, published in 1798 as *Essay on the Theory of Numbers*. It went through several editions, the final one appearing in 1830 as *Theory of Numbers*. In the first and second editions Legendre reproduces Euler's proofs of FLT for exponents 3 and 4. Then, in an 1825 supplement to the second edition, he adds some work of his own, including a proof for exponent 5. Legendre's contribution to this case consists in completing a partial proof given by the young German mathematician Lejeune Dirichlet (1805–1859) in the same year.

By the time the second edition of *Theory of Numbers* appeared in 1808 it had been made utterly obsolete by an amazing work by the young German mathematician Carl Friedrich Gauss (1777–1855), who in 1801 published a book entitled *Disquisitiones Arithmeticae* (Arithmetical Investigations), which laid much of the foundation of modern number theory. It contained proofs of a number of results such as the *Quadratic Reciprocity Theorem*, one of the fundamental facts about prime numbers, which had been conjectured by Euler and for which Legendre had provided an incorrect proof. In it, Gauss developed the theory of congruence arithmetic, which is still in use today. (See the Appendix to this chapter.) The *Disquisitiones* finally established number theory as a mathematical theory with a coherent body of results and techniques. And gradually some of the greatest mathematical minds of the nineteenth century fell under the spell of the new subject. Its prosperity from then on was assured, with a plethora of new results and methods coming forth continuously throughout the second half of the nineteenth and the twentieth century.

Gauss's view on FLT is summarized in a letter to his colleague W. Olbers, dated March 21, 1816:

> I do admit that the Fermat Theorem as an isolated result is of little interest to me, since it is easy to postulate a lot of such theorems, which one can neither prove nor refute. Nonetheless, it has caused me to return to some old ideas for a *great* extension of higher arithmetic. Of course, this theory is one of those things where one cannot presuppose to what extent one will succeed in reaching goals looming in the far distance. A lucky star must also preside, and my situation as well as much detracting business do not allow me to indulge in such meditations as during the lucky years 1796–1798, when I formed the main parts of my Disquisitiones Arithmeticae.

> Alas, I am convinced, that if *luck* contributes more than I am allowed to hope for, and I succeed in some of the main steps in that theory, then the Fermat theorem will appear in it as one of the least interesting corollaries [151, p. 629].

But luck did not favor Gauss that time, and he never returned to serious number-theoretic investigations. Nonetheless, the *Disquisitiones* and its congruence arithmetic immediately inspired a whole new line of attack on FLT in its full generality, rather than one exponent at a time. In Paris, the young Sophie Germain (1776–1831) devoured Gauss's book, after having studied Legendre's *Essay*. She immediately perceived a way to use congruence methods to get at a general proof of FLT, and devoted much of her life to this ultimately unsuccessful effort. But she did succeed in proving the first general result about FLT, and her approach to the problem was pursued quite successfully by many researchers even into the 1980s.

The third original source in this chapter is the only result commonly attributed to her, known as Sophie Germain's Theorem. Germain never published any of her work on FLT, and the only published reference to this theorem consists of a footnote in the above-mentioned supplement to Legendre's second edition of his *Essay*, which deals with FLT. Here we present an excerpt from her hand-written manuscripts, archived in the Bibliothèque Nationale, in Paris, and from unpublished correspondence with Gauss.

Fermat had already observed that it was sufficient to prove FLT for exponent four and for odd prime exponents p (Exercise 4.10). Germain proved that for such p, if there were nonzero numbers x, y, z such that

$$x^p + y^p = z^p,$$

and in addition an auxiliary prime q satisfying certain properties, then p^2 would have to divide one of the numbers x, y, z. She then proceeded to develop an algorithm to find such auxiliary primes q, and used it successfully for all primes less than 100. Her method for generating auxiliary primes is easily applied to higher prime exponents, as was done by Legendre, who extended the list to include all prime exponents up to 197. Consequently, for any prime exponent less than 197, any solutions to the Fermat equation would have to contain one number that is divisible by the exponent. This result is the origin of a case distinction that has been made ever since. Solutions to the Fermat equation such that xyz is not divisible by the exponent are referred to as Case I solutions, the others as Case II.

While there were a number of women who played a significant role in the development of mathematics before the time of Germain, such as Hypatia, in classical Greece, or Maria Gaetana Agnesi, during the Renaissance, Sophie Germain was the first woman in history who we know produced significant original mathematical research, working in both number theory and mathematical physics [79, 132]. An excellent biography of Germain is [23].

By the middle of the nineteenth century, sophisticated new methods were being applied to FLT. The chapter ends with a letter from the German number theorist Ernst Kummer (1810–1893) to Joseph Liouville, in Paris, in May 1847. The letter addresses the failure of unique factorization into primes of certain complex numbers, similar to that of integers into products of prime numbers. Several proposed proofs of FLT had tacitly assumed that such unique factorization held in great generality, and Kummer pointed out that he had obtained results to the contrary. Kummer's study of this problem through entirely new methods was a radical departure from the work of his predecessors and marks one of the beginnings of algebraic number theory. His main positive contribution to FLT was a proof that it was true for certain prime exponents called *regular*. In particular, all primes less than 100, except for 37, 59, 67, are regular.

In the following century and a half, the number-theoretic world inched ever closer to a complete proof of the theorem. Good surveys can be found in [47, 137]. Many supporting partial results were achieved, such as the result proved by A. Wieferich in 1909 that if there is a Case I solution for an exponent p, then p must satisfy the congruence

$$2^{p-1} \equiv 1 \pmod{p^2}.$$

In 1976, it was shown that FLT is true for all prime exponents less than 125,000. In 1992 this was extended to 4,000,000. Things really started to get exciting in 1983, when the German mathematician Gerd Faltings proved the so-called Mordell Conjecture, a result in algebraic geometry that implies that for a given $n \geq 4$, the equation $x^n + y^n = z^n$ has at most finitely many pairwise relatively prime integer solutions, that is, where x, y, z have no common divisors. Algebraic geometry concerns itself with the study of the solution set to a system of polynomial equations, such as

$$y - x^2 = 0,$$
$$z - x^3 = 0.$$

The solution set to such a system depends, of course, on the type of numbers one allows, such as rational or real numbers, in which cases one may as well rewrite the Fermat equation by dividing both sides by z to get the equation $x^p + y^p = 1$. Given a rational solution to this equation, we obtain an integer solution to the Fermat equation by clearing denominators. In the plane, the equation $x^p + y^p = 1$ has as solution set a curve. And FLT is equivalent to the assertion that this curve contains no points whose coordinates are rational numbers. Thus, if one views FLT as a problem in algebraic geometry, one can bring to bear on it many tools from this subject, in addition to number-theoretic ones. The Mordell conjecture asserted that certain types of curves, such as the curves $x^n + y^n = 1$, for $n \geq 5$, have only finitely many rational points, and thus, in particular, the Fermat equation has only finitely many integer solutions in which the numbers are pairwise relatively prime.

Of course, Faltings' result was still far from FLT. But it did allow A. Granville and D. Heath-Brown to prove in 1985 that FLT holds for "most" exponents n, in the sense that as n increases, the probability that FLT fails for n approaches zero. Now the pace of results was quickening. Several conjectures in number theory were made, each of which would imply FLT if found true. One of those, the so-called Taniyama–Shimura conjecture, named after two Japanese mathematicians, pertained to elliptic curves with rational coefficients, which are curves defined by an equation of the type

$$y^2 = ax^3 + bx^2 + cx + d,$$

such that the coefficients a, b, c, d are rational numbers with $a \neq 0$ and the polynomial on the right side of the equation has distinct roots. The conjecture asserted that such curves necessarily had to be *modular*. (It is beyond the scope of this book to discuss this property in detail, and the interested reader is advised to consult the excellent article [33], also [117, 138, 158].) The reason that this conjecture, if proven true, would close the chapter on FLT was a result from the early eighties due to the German mathematician Gerhard Frey. He showed that a nontrivial solution to FLT would allow the construction of a certain elliptic curve with special properties, which he thought would prevent it from being modular. These elliptic curves, now called Frey curves, are constructed as follows. Given a nontrivial solution $a^p + b^p = c^p$ for a particular exponent $p \geq 5$, the associated Frey curve is

$$y^2 = x(x - a^p)(x + b^p).$$

In 1986, Ken Ribet, from Berkeley, completed the last step needed to confirm Frey's intuition. Thus, all that was needed to complete the proof of FLT was a proof of the Taniyama–Shimura conjecture.

In June of 1993, Princeton mathematician Andrew Wiles gave a series of three lectures at the Isaac Newton Institute in Cambridge, England, in which he outlined a proof of the Taniyama–Shimura conjecture for a certain class of elliptic curves, including the Frey curves. Thus, a proof of Fermat's Last Theorem seemed complete, after more than 300 years. Wiles produced a lengthy manuscript with the details of his extremely intricate and difficult arguments, which he submitted to the scrutiny of several experts in the field. After a lengthy silence from the refereeing committee, rumors of a supposed gap in the proof alarmed the mathematics community, which had already witnessed a false sense of triumph some years earlier, when a purported proof turned out to be incomplete. Indeed, it was becoming clear that Wiles's proof contained a gap as well. Fortunately, in September of 1994, Wiles and Cambridge (UK) mathematician Richard Taylor managed to circumvent this gap and produce a complete proof, which has since been scrutinized very carefully and found to be complete and correct.

On June 27, 1997, Andrew Wiles received the Wolfskehl Prize in Göttingen, Germany. This prize had been established by the German mathematician Paul Wolfskehl (1856–1906), who had become fascinated by the problem through the lectures and papers of Ernst Kummer. The first person to give a correct proof of Fermat's Last Theorem or a necessary and sufficient criterion for those exponents for which the Fermat equation is unsolvable in positive integers was to receive 100,000 German marks. (When it was awarded to Wiles, it was valued at approximately \$43,000.) The prize triggered an initial deluge of incorrect proofs. (For details on the prize see [7, pp. 1294–1303].)

One of the crowning achievements of twentieth-century mathematics, the proof of Wiles and Taylor brings to an end an odyssey spanning almost four centuries. At the same time, the advances in understanding that made the proof possible have spawned fascinating new questions that will continue to drive mathematics in the future, just as Fermat's Last Theorem did in the past. We are extremely fortunate to live during one of the most exciting times in the whole history of mathematics,

PHOTO 4.6. Wiles.

and all indications are that we are in for a thrilling number-theoretic ride in the future.

Exercise 4.1: Look up and understand the proof of Euclid's proposition about perfect numbers in his *Elements*. Use it to find as many perfect numbers as you can.

Exercise 4.2: Explain how to translate Fermat's statement of Fermat's Little Theorem into the modern version.

Exercise 4.3: Show that if n is a positive integer, and $2^n - 1$ is prime, then n is also prime. Hint: Prove the equality

$$(2^{ab} - 1)/(2^a - 1) = 2^{a(b-1)} + 2^{a(b-2)} + \cdots + 1.$$

Exercise 4.4: Use Fermat's Little Theorem to prove the corollary that if p is an odd prime and q is a prime that divides $2^p - 1$, then q is of the form $2kp + 1$ for some integer k.

Exercise 4.5: Suppose that $a^n + 1$ is prime. Show that a must be even and n must be a power of 2. (Hint: Prove that if $n = 2^k m$, with $m > 1$ odd, then

$$(a^{2^k m} + 1)/(a^{2^k} + 1) = a^{2^k(m-1)} - a^{2^k(m-2)} + a^{2^k(m-3)} - \cdots + 1.)$$

Exercise 4.6: What integer Pythagorean triple results from Diophantus's solution to the problem of dividing a given square into two squares?

Exercise 4.7: A *primitive* Pythagorean triple is one in which any two of the three numbers are relatively prime. Show that every multiple of a Pythagorean triple is again a Pythagorean triple, and that every Pythagorean triple is a multiple of a primitive one.

Exercise 4.8: Show that the sum of two odd squares is never a square, and use this fact to conclude that all Pythagorean triples have an even leg.

Exercise 4.9: Look up the Euclidean algorithm and use it to decide whether a Pythagorean triple is primitive or not.

Exercise 4.10: Show that FLT is true for all exponents n if it is true for $n = 4$ and all odd prime numbers n.

4.2   Euclid's Classification of Pythagorean Triples

A triple of positive integers (x, y, z) is called a *Pythagorean triple* if the integers satisfy the equation $x^2 + y^2 = z^2$. Such a triple is called *primitive* if x, y, z have no common factor. For instance, $(3, 4, 5)$ and $(5, 12, 13)$ are primitive triples, whereas $(6, 8, 10)$ is not primitive, but is a Pythagorean triple. The significance of primitive triples is that "multiples" of primitive ones account for all triples (see Exercise 4.7 in the previous section). The problem of finding Pythagorean triples occupied the minds of mathematicians as far back as the Babylonian civilization. Analysis of cuneiform clay tablets shows that the Babylonians were in possession of a systematic method for producing Pythagorean triples [127, pp. 36 ff.].

For instance, the tablet catalogued as Plimpton 322 in Columbia University's Plimpton Collection, dating from 1900–1600 B.C.E., contains a list of fifteen Pythagorean triples as large as $(12709, 13500, 18541)$. (Is this triple primitive?) For a detailed discussion of this tablet see, e.g., [64], [129]. There is reason to believe that the Babylonians might even have known the complete solution to the problem [131, pp. 175–79]. Other civilizations, such as those of China and India, also have studied the problem [93]. Clearly, Pythagorean triples are related to geometry via the Pythagorean Theorem, as a Pythagorean triple corresponds to a right triangle with integer sides.

The Pythagoreans, after whom the theorem is named, were an ancient Greek school that flourished around the sixth century B.C.E. Aristotle says that they "applied themselves to the study of mathematics, and were the first to advance that science; insomuch that, having been brought up in it, they thought that its principles must be the principles of all existing things" [85, p. 36]. Their motto is said to have been "all is number" [20, p. 54]. The particular interest of the Pythagoreans in relationships between whole numbers naturally led to the investigation of right triangles with integral sides. Proclus, a later commentator, who taught during the

PHOTO 4.7. Plimpton 322.

fifth century C.E. at the Neo-Platonic Academy in Athens, credits the Pythagoreans with the formula

$$(2n + 1, 2n^2 + 2n, 2n^2 + 2n + 1),$$

for any positive integer n, generating triples in which the longer leg and hypotenuse differ by one. He also describes another formula,

$$(2n, n^2 - 1, n^2 + 1)$$

for all $n \geq 2$, attributed to Plato [58, pp. 212–13], [85, p. 47], [134, p. 340] (Exercise 4.18).

The first formula giving a complete classification of all triples appears in Euclid's *Elements*:

> To find two square numbers such that their sum is also square [51, vol. 3, p. 63].

This is the first text that we will examine.

Very little is known about Euclid's life. The most that can be said with any certainty is that he lived about 300 B.C.E. and taught at the University in Alexandria. He probably attended Plato's Academy in Athens, which was at the time the center of Greek mathematics, and he was invited to Alexandria when the great library and museum were being set up at the direction of Ptolemy Soter. Under the rule of Ptolemy, Alexandria became a world center both in commerce and scholarship; Euclid was a founder of the great school of mathematics there. Proclus tells us that Euclid "put together the *Elements*, collecting many of Eudoxus's theorems, perfecting many of Theaetetus's, and also bringing to irrefragable demonstration the things which were only somewhat loosely proved by his predecessors" [85, p. 202]. There is an extensive entry on Euclid in [42], which is highly recommended. The *Elements* ended up as the all-time number one mathematical bestseller, and was used as a textbook in a number of countries even into this century. We quote from

PHOTO 4.8. Pythagoras.

the preface of one such textbook [170] used in England during the second half of the nineteenth century:

> In England, the text-book of Geometry consists of the *Elements* of Euclid; for nearly every official programme of instruction or examination explicitly includes some portion of this work. Numerous attempts have been made to find an appropriate substitute for the *Elements* of Euclid; but such attempts, fortunately, have hitherto been made in vain. The advantages attending a common standard of reference in such an important subject, can hardly be overestimated; and it is extremely improbable, if Euclid were once abandoned, that any agreement would exist as to the author who should replace him.

The *Elements* consists of thirteen books, which deal with a variety of different subjects. No manuscripts from Euclid's time are known, and the versions of the *Elements* in use today are translations of Arabic, Greek, and Latin versions of various time periods. The first printed edition of the *Elements* was published at Venice in 1482. It was the first printed mathematics book of any importance. For more details on the various translations see [85, Ch. IX]. For a more detailed discussion of the structure of the *Elements* see the geometry chapter.

PHOTO 4.9. First printed edition of Euclid, 1482.

Euclid's solution to the problem of the classification of all Pythagorean triples appears as Lemma 1 in Book X, before Proposition 29.[2] Book X concerns itself primarily with the theory and application of irrationals, and is the first major treatise we have on this subject. Pythagorean triples are used in Proposition 29 to produce examples of irrationals with certain properties.

The reader should be cautioned that the following text is rather subtle and requires very careful reading. It is followed by an extensive mathematical commentary. In order to understand it, one needs to know what *similar plane numbers* are, the central concept in Euclid's argument. First of all, a *plane* number is simply a formal product of two numbers, thought of as the sides of a rectangle, of

---

[2]Most of the statements in the *Elements* are called Propositions, followed by a proof. Then there are other statements, referred to as "Lemma." These are simply auxiliary results that are needed in the proof of a proposition, but that are stated and proved separately, so as not to interrupt or unduly lengthen the proof. They may be proved before or after the proposition in which they are used.

which the plane number itself gives the area. Two plane numbers are *similar* if their respective sides are proportional with the same rational proportionality factor. Geometrically, the two plane numbers represent similar rectangles. For example, the two plane numbers 6 · 10 and 9 · 15 are similar.

Euclid, from

Elements

LEMMA 1
(before Proposition 29 in Book X)

To find two square numbers such that their sum is also a square.

Let two numbers AB, BC be set out, and let them be either both even or both odd.

Then since, whether an even number is subtracted from an even number, or an odd number from an odd number, the remainder is even [IX. 24, 26], therefore the remainder AC is even.

Let AC be bisected at D.

Let AB, BC also be either similar plane numbers, or square numbers, which are themselves also similar plane numbers.

Now the product of AB, BC together with the square on CD is equal to the square on BD [II. 6].

And the product of AB, BC is square, inasmuch as it was proved that, if two similar plane numbers by multiplying one another make some number the product is square [IX. 1].

Therefore two square numbers, the product of AB, BC, and the square on CD, have been found which, when added together, make the square on BD.

And it is manifest that two square numbers, the square on BD and the square on CD, have again been found such that their difference, the product of AB, BC, is a square, whenever AB, BC are similar plane numbers.

But when they are not similar plane numbers, two square numbers, the square on BD and the square on DC, have been found such that their difference, the product of AB, BC, is not square.

Q. E. D.

In order to translate Euclid's argument into modern algebraic notation, let $AB = u$, $BC = v$, $CD = d$, $BD = f$. From [II.6][3] Euclid obtains the equation

$$uv + d^2 = f^2,$$

---

[3]This states: "If a straight line be bisected and a straight line be added to it in a straight line, the rectangle contained by the whole with the added straight line and the added straight

which alternatively follows via

$$uv = (f + d)(f - d) = f^2 - d^2 ,$$

since D bisects AC and thus $AD = CD = d$.

The remainder of Euclid's argument involves similar plane numbers. One can prove that two numbers are similar plane numbers if and only if their ratio is the square of a rational number (see the lemma at the end of the section). Euclid continues his analysis by noting that uv is a square if and only if u and v are similar plane numbers, which follows from the previous sentence. So far he has given a procedure that beginning with two similar plane numbers of the same parity constructs a Pythagorean triple. (The parity of a number is simply whether it is even or odd. Thus 4 is of even parity. That two numbers are of the same parity means that they are either both even or both odd.) (Observe that his plane numbers must be unequal; why?)

But in fact all Pythagorean triples arise in this way, as he somewhat cryptically explains in the last paragraph. Start with a difference $f^2 - d^2$, and factor it as uv, where $u = f + d$ and $v = f - d$. Since this u and v are clearly of the same parity, Euclid's construction can proceed using these numbers. Notice that this will reproduce the original d and f with which we started. The result is the equation $uv = f^2 - d^2$ as above, but as Euclid says in his last paragraph, this difference of squares, which is uv, will only itself be a square if u and v were similar plane numbers to begin with. Thus Euclid has set up a one-to-one correspondence between differences of squares and pairs of unequal numbers of the same parity, in such a way that differences of squares that are themselves square correspond to pairs that are similar plane numbers. Thus he has classified all Pythagorean triples in terms of similar plane numbers; so the problem of finding all Pythagorean triples reduces to that of finding all pairs of similar plane numbers of the same parity (Exercises 4.12–4.13).

Now we can see how to find the formulas of the Pythagoreans and Plato that we mentioned. Since

$$f = \frac{u + v}{2} \text{ and } d = \frac{u - v}{2},$$

we have

$$\left(\frac{u + v}{2}\right)^2 - \left(\frac{u - v}{2}\right)^2 = uv.$$

The simplest way to choose u, v such that their product is a square is to let $u = k^2$ and $v = 1$. For f and d to be integers, we must have k odd, thus leading to the formula of the Pythagoreans. Letting $u = 2k^2$ and $v = 2$ gives the Platonic formula [51, Vol. 1, p. 385].

---

line together with the square on the half is equal to the square on the straight line made up of the half and the added straight line."

From Euclid's classification of Pythagorean triples we can derive the following classification of all primitive Pythagorean triples, which we will need for our next source. Suppose that the triple (d, e, f) is primitive. One of the numbers d, e has to be even (this follows from Exercise 4.8 in the Introduction); let us assume that it is e, by interchanging d and e if necessary. Now, as noted above, we can apply Euclid's method to this triple, obtaining

$$f^2 - d^2 = e^2 = uv,$$

with u, v similar plane numbers of the same parity. Thus u and v are both even, and it follows from the lemma below that they are of the form $u = mp^2$, $v = mq^2$. A common divisor of u and v also divides $u + v = 2f$ and $u - v = 2d$. But since d and f are relatively prime, that is, their greatest common divisor is 1, it follows that the greatest common divisor of u and v is 1 or 2. Together with the fact that u and v are even, this implies that $m = 2$, and that p and q are relatively prime. Thus, the triple is of the form

$$d = p^2 - q^2, \qquad e = 2pq, \qquad f = p^2 + q^2,$$

and moreover, p and q have different parity, since d is odd by primitivity of the triple.

To summarize, any primitive triple has this form (by interchanging d and e if necessary), where p is greater than q, and p and q are relatively prime and of different parity. Furthermore, any such choice of p and q results in a primitive triple (check this).

Lemma: *Two plane numbers are similar if and only if their ratio is the square of a rational number.*

Proof. Suppose that u and v are similar plane numbers. Then $u = xy$ and $v = zw$ for some numbers (that is, positive integers) x, y, z, w, which are the sides of u and v, respectively. For these sides to be proportional means that $x = ry$ and $z = rw$ for some rational number r. So

$$\frac{u}{v} = \frac{xy}{zw} = \frac{ry^2}{rw^2} = \left(\frac{y}{w}\right)^2,$$

and y/w is clearly rational. Conversely, suppose $u/v = (p/q)^2$, where we may assume that p and q are relatively prime. So

$$u = \frac{p^2}{q^2}v.$$

As u is an integer, we know that q^2 divides $p^2 v$. Since p and q are relatively prime, q^2 must divide v; that is, $v = mq^2$ for some positive integer m. Similarly, p^2 must divide u, and it is easy to see that in fact $u = mp^2$. If we choose p and mp to be the sides of u, and q and mq to be the sides of v, we see that u and v are similar plane numbers.

Exercise 4.11: Prove that there are only a finite number of Pythagorean triples with a fixed leg.

Exercise 4.12: Show that there are exactly two pairs of similar plane numbers of same parity that will produce a given Pythagorean triple using Euclid's construction.

Exercise 4.13: Enumerate (that is, give a recipe for listing) all pairs of similar plane numbers with same parity. (Remember to show that your list is complete.) Use your list to enumerate all Pythagorean triples.

Exercise 4.14: Enumerate all primitive Pythagorean triples.

Exercise 4.15: Derive a general formula from Diophantus's treatment of Pythagorean triples [86], and compare it with Euclid's.

Exercise 4.16: Give a geometric demonstration of Euclid's statement that "the product of AB, BC together with the square on CD is equal to the square on BD"; that is,

$$AB \cdot BC + (CD)^2 = (BD)^2.$$

[Hint: Recall that the product of two numbers may be represented geometrically as the area of a rectangle with those numbers as the lengths of its sides.]

Exercise 4.17: Enumerate as many Pythagorean triples as you can that contain the number 1378.

Exercise 4.18: In the two formulas for Pythagorean triples, ascribed to the Pythagoreans and Plato, respectively, which triples, if any, appear on both lists? Are the triples generated by these lists primitive? Which are, and which are not? Are there an infinite number of primitive triples? Do the primitive triples in the above two lists exhaust all possible primitive triples?

4.3 Euler's Solution for Exponent Four

Without doubt Leonhard Euler is one of the world's mathematical giants, whose work profoundly transformed mathematics. He made extensive contributions to many mathematical subjects, including number theory, and was so prolific that the publication of his collected works, begun in 1911, is still underway, and is expected to fill more than 100 large volumes.

Born in Basel, Switzerland, in 1707, Euler's mathematical career spanned almost the whole eighteenth century, and he was at the heart of all its great accomplishments. His father, a Protestant minister, was interested in mathematics and had attended Jakob Bernoulli's lectures at the University of Basel. He was responsible for his son's earliest education, including mathematics. Later, Euler attended the Gymnasium in Basel, a high school that did not provide instruction in mathematics, however. At fourteen, Euler entered the University of Basel, where Johann

PHOTO 4.10. Euler.

Bernoulli (1667–1748) had succeeded his brother Jakob in the chair in mathematics. Though Bernoulli declined to give Euler private lessons (and Bernoulli's public lectures at the university were limited to elementary mathematics), he was willing to help Euler with difficulties in the mathematical texts that Euler studied on his own.

Euler received a degree in philosophy and joined the department of theology in 1723, but his studies in theology, Greek, and Hebrew suffered from his devotion to mathematics. Eventually he gave up the idea of becoming a minister. In autumn of 1725, Johann Bernoulli's sons Nikolaus and Daniel went to Russia to join the newly organized St. Petersburg Academy of Sciences; at their behest, the following year the Academy invited Euler to serve as adjunct of physiology, the only position available at the time. Euler accepted, arriving in St. Petersburg in May of 1727. In spite of having been invited to study physiology, soon after his arrival he was given the opportunity to work in his true field of mathematics. During fourteen years in St. Petersburg, Euler published fifty-five works, making brilliant discoveries in such fields as analysis, number theory, and mechanics.

In 1740, Euler was invited to join the Berlin Academy of Sciences and accepted, since the political situation in St. Petersburg had deteriorated by that time. During his tenure in Berlin, he remained an active member of the St. Petersburg Academy as well, publishing prolifically in both Academies. In 1766, Euler returned to St. Petersburg, where he was to remain for the rest of his life. Though he went blind shortly after his return, he was able to continue his work with the aid of assistants; indeed, he actually increased his output.

From Euler's extensive work in number theory, we will examine his proof of Fermat's Last Theorem for exponent four. In a letter to Christian Goldbach in 1753, Euler wrote:

> There is another very nice theorem in Fermat, whose proof he claims to have found. Namely, when considering the Diophantine problem of finding two squares whose sum is a square, he says that it is impossible to find two cubes whose sum is a cube, and two biquadratics whose sum is biquadratic, and more generally, that the formula $a^n + b^n = c^n$ is always impossible when $n > 2$. I have now indeed found proofs that $a^3 + b^3 \neq c^3$ and $a^4 + b^4 \neq c^4$, where $\neq$ indicates the impossibility of equality. But the proofs for these cases are so different from each other, that I do not see any possibility of deriving therefrom a general proof for $a^n + b^n \neq c^n$ for $n > 2$. Indeed, one sees rather clearly by extension that the larger n is the more impossible the formula must be. In fact I have not even been able to prove that a sum of two fifth powers cannot be a fifth power. It appears that this proof [for $n = 3$ and 4] rests on a lucky idea, and if one does not stumble upon it, all effort is in vain. But since the equation $aa + bb = cc$ is possible, so also is $a^3 + b^3 + c^3 = d^3$ possible. From this it seems to follow that $a^4 + b^4 + c^4 + d^4 = e^4$ is also possible, but I have yet to find a single occurrence of it. One can however give five biquadrates whose sum is a biquadrate [54, pp. 614–618].

In fact, Euler proved something slightly stronger than FLT for exponent four, namely that

$$a^4 + b^4 = c^2$$

has no positive integer solutions. FLT for exponent four is an immediate consequence of this theorem, since any solution to Fermat's equation $x^4 + y^4 = z^4$ would immediately provide a solution to Euler's equation by writing $z^4 = (z^2)^2$.

Euler gave two proofs for exponent four, first in a 1738 paper entitled *Theorematum quorundam arithmeticorum demonstrationes* [52], and later in his 1770 book *Vollständige Anleitung zur Algebra* (Elements of Algebra) [53, pp. 436–439]. Following is the latter proof.

Leonhard Euler, from
Elements of Algebra

ON SOME FORMULAS OF THE FORM $ax^4 + by^4$, WHICH ARE NOT REDUCIBLE TO A SQUARE

202.

Much effort has been expended to find two biquadrates whose sum or difference is a square number; alas, all effort was in vain. Finally, a proof was even found that neither the formula $x^4 + y^4$ nor $x^4 - y^4$ could ever become a square, with the exception of two cases, namely when either $x = 0$ or $y = 0$ in the first case, and $y = 0$ or $y = x$ in the second; and in both of these cases the situation is obvious. But that it should be impossible in all other cases is all the more curious, because there are infinitely many solutions if one considers only mere squares.

203.

In order to properly present the proof, one must first of all remark that one may assume that the two numbers x and y may be regarded as relatively prime. Because, suppose that they have a common divisor d, then one could write $x = dp$ and $y = dq$. If our formulas $d^4 p^4 + d^4 q^4$ and $d^4 p^4 - d^4 q^4$ were squares, then they would have to remain squares after division by d^4. But then the formulas $p^4 + q^4$ and $p^4 - q^4$ would also be squares, where now the numbers p and q do not have any further common divisors. It is therefore sufficient to prove that these formulas cannot become squares in the case where x and y are relatively prime, and the proof subsequently extends to all cases where x and y have common divisors.

204.

We will begin with the sum of two biquadrates, namely the formula $x^4 + y^4$, where we can assume that x and y are relatively prime. The proof that $x^4 + y^4$ cannot be a square, except in the above mentioned cases, now proceeds as follows.

If someone wanted to deny the theorem, then he would have to claim that there are values for x and y which make $x^4 + y^4$ a square, regardless of how large they might be, since surely there are none among the small numbers.

But one can clearly show that if there would exist such values for x and y among the largest numbers, then one could conclude that there would be such values also among smaller numbers, and from these there would be smaller numbers still, etc. But since there are no such values among small numbers, except for the two mentioned above, which don't lead to any others, one can surely conclude that there can be no such values for x and y in larger, and even the largest, numbers. And one can prove in the same manner the theorem about the difference of two biquadrates $x^4 - y^4$, as we are going to show promptly.

205.

To show first that $x^4 + y^4$ cannot be a square, except for the two cases, which are clear, one needs to observe the following theorems.

I. Let us assume that the numbers x and y are relatively prime, or have no common divisor. Then either both are odd, or one is odd and the other is even.

II. But not both can be odd, because the sum of two odd squares can never be a square: an odd square is always of the form $4n + 1$, and therefore a sum of two odd squares has the form $4n + 2$, which is divisible by 2 but not by 4, and can thus not be a square. But the same is true for two odd biquadrates.

III. If, therefore, $x^4 + y^4$ would be a square, then one would be even and the other odd. But we have seen above that if the sum of two squares should be a square, then the root of one can be expressed as $pp - qq$, the root of the other as $2pq$. From this it follows that we would have to have $xx = pp - qq$ and $yy = 2pq$, so that $x^4 + y^4 = (pp + qq)^2$.

IV. So here y would be even, but x odd. But since $xx = pp - qq$, one of the numbers p and q has to be even and the other odd. But the former, p, cannot be even, since then $pp - qq$, as a number of the form $4n - 1$ or $4n + 3$, can never become a square. Therefore, p has to be odd, but q even, and it is assumed that they are relatively prime.

V. But since $pp - qq$ should be a square, namely equal to xx, this happens when $p = rr + ss$ and $q = 2rs$, as we have seen above. Then $xx = (rr - ss)^2$, and therefore also $x = rr - ss$.

VI. But yy also must be a square. Since $yy = 2pq$, we now get $yy = 4rs(rr + ss)$, which must be a square. Consequently, $rs(rr + ss)$ must also be a square, where r and s are relatively prime numbers, so that the three factors r, s and $rr + ss$ also cannot have a common factor.

VII. But if a product with several factors, which are relatively prime, is supposed to be a square, then each factor on its own has to be a square, so that one can set $r = tt$ and $s = uu$. Then $t^4 + u^4$ must also be a square. So if, therefore, $x^4 + y^4$ were a square, then $t^4 + u^4$, a sum of two biquadrates, would also be a square. One needs to remark that, since $xx = (t^4 - u^4)^2$ and $yy = 4ttuu(t^4 + u^4)$, the numbers t and u would obviously be much smaller than the numbers x and y. Since x and y are determined even by the fourth powers of t and u, they would undoubtedly have to be much bigger.

VIII. Therefore, if there were two biquadrates x^4 and y^4 among the largest numbers, whose sum was a square, then one could derive a sum of two smaller biquadrates, which would also be a square. And from these one could once again derive a smaller such sum, etc., until one would finally arrive at very small numbers: but since such a sum is impossible in small numbers, it follows obviously that there is no such thing among the largest numbers either.

IX. One might object here that there really are such values among the small numbers, as remarked in the beginning, namely where one biquadrate becomes zero. But one definitely never arrives at this case, when one regresses from the largest numbers to smaller ones. Since, if in the smaller sum $t^4 + u^4$, either $t = 0$ or $u = 0$, then in the larger sum one would necessarily have $yy = 0$. But we are not considering that case here.

In order to discuss Euler's method of proof, let us call a positive integer solution (x, y, z) of the equation $x^4 + y^4 = z^2$ an *Euler triple*. Euler's goal is to show that no Euler triples exist. Not surprisingly, he uses Fermat's method of infinite descent (Exercise 4.19). In order to apply this method, Euler describes a procedure, or algorithm, which takes as its input any Euler triple and produces another Euler triple as its output; moreover, the numbers in the output triple are smaller than those in the input. This algorithm lies at the heart of applying the method of infinite descent, since the existence of such an algorithm proves the theorem that there are no Euler triples, as follows. The algorithm Euler describes can be thought of as a machine whose input is an Euler triple and which produces another, smaller, Euler triple as output. The output could then be fed in again as input to the machine, thereby producing smaller and smaller Euler triples (of positive integers) ad infinitum. Since this is clearly impossible, it must not be possible to start up the machine to begin with. In other words, no Euler triple can exist with which to start the machine.

The algorithm begins by reducing the given Euler triple to a primitive one, that is, one in which the three numbers are pairwise relatively prime, by dividing out any common factors. Notice that this reduction does not increase the numbers in the triple. Since (x^2, y^2, z) is a primitive Pythagorean triple, Euler can now use the classification we derived from Euclid at the end of the previous section to express x^2 and y^2 in terms of suitably chosen numbers p and q. Since (x, q, p) is another primitive Pythagorean triple, it follows that p is odd, as Euler claims.

Euler can now use Euclid's classification once again, this time writing the triple (x, q, p) in terms of new numbers r and s. After showing that r and s are the squares of numbers u and t in a new Euler triple, it suffices to show that u and t are smaller than the input numbers x and y. For instance,

$$x = r^2 - s^2 = (r - s)(r + s) = (t^2 - u^2)(t^2 + u^2),$$

and thus u and t are both less than x. Similarly, one shows that u and t are both less than y (Exercises 4.20–4.21).

As mentioned in the Introduction, Euler also gave a proof of FLT for exponent three, which is incomplete at one point, however. Again, Euler relies on the method of infinite descent. He first reduces to the case of a solution to $x^3 + y^3 = z^3$, in which the three integers are pairwise relatively prime. Then he argues that exactly one of the three integers is even. Suppose first that z is even. Then $x + y = 2p$ and $x - y = 2q$ are both even. If one now expresses x and y in terms of p and q in the factorization

$$x^3 + y^3 = (x + y)(x^2 - xy + y^2),$$

then one can conclude from the resulting equation that p and q have to have opposite parity, are relatively prime, and $2p(p^2 + 3q^2)$ has to be a cube. The conclusion applies also when z is assumed to be odd.

The argument now continues by considering two cases, depending on whether $2p$ and $p^2 + 3q^2$ are relatively prime or not. In the first case the only way the product can be a cube is for each factor to be a cube. Then Euler argues that if these expressions are cubes, then there exist integers a and b such that

$$p = a^3 - 9ab^2 \quad \text{and} \quad q = 3a^2b - 3b^3.$$

Consequently, $p^2 + 3q^2 = (a^2 + 3b^2)^3$. Factoring the above expressions for p and q, one obtains that

$$2p = 2a(a - 3b)(a + 3b)$$

has to be a cube. After showing that the three factors are necessarily relatively prime, he concludes that each one of them has to be a cube itself, say

$$2a = \alpha^3, a - 3b = \beta^3, a + 3b = \gamma^3.$$

But then $\beta^3 + \gamma^3 = \alpha^3$, and we have a solution to the Fermat equation in smaller numbers than x, y, and z. The second case is a variation of the first, after showing that the only possible common divisor of $2p$ and $p^2 + 3q^2$ is 3. This proves the theorem for exponent three (Exercise 4.25).

Unfortunately, Euler's argument is not quite complete. He fails to show that $p^2 + 3q^2$ can be a cube only in the above way. An argument for this using reasoning employed by Euler elsewhere can be found in [47, pp. 52–54].

Exercise 4.19: Look up mathematical induction and proof by contradiction. (One appropriate reference is [161].) What property of the natural numbers do both mathematical induction and the method of infinite descent exploit?

Exercise 4.20: What are some weaker conditions on the size of u and t in relation to x and y that would still make the method of infinite descent work?

Exercise 4.21: Adapt Euler's proof to show that $x^4 - y^4$ cannot be a square.

Exercise 4.22: Show that if a and b are relatively prime numbers and ab is a square, then both a and b are squares. Hint: Use the Fundamental Theorem of Arithmetic, which says that any positive integer can be factored into a product of prime numbers, and the prime factors are unique up to their order in the product.

Exercise 4.23: Read about Christian Goldbach and discuss his influence on Euler's number-theoretic work.

Exercise 4.24: Lagrange wrote an appendix to Euler's *Elements of Algebra*. Look up the appendix in Euler's *Opera Omnia* and discuss its contents.

Exercise 4.25: Use [47] to complete the details of Euler's proof for exponent three.

4.4   Germain's General Approach

Sophie Germain was the first mathematician to make progress with a general approach toward proving Fermat's Last Theorem. Born in Paris into a family of the French middle class, she was able to overcome the obstacles to her education from society, and the disapproval of her parents, by using her father's extensive library, sometimes even clandestinely, to educate herself at home.

Higher education in mathematics was virtually nonexistent in France until the founding of the Ecole Polytechnique in 1795 as part of educational reforms after the Revolution, with its primary mission the education of military engineers and civil servants. It was more ambitious than the institutions that preceded it,

PHOTO 4.11. Germain.

eventually setting the standard for technical education throughout the Western world and making Paris the world center of mathematical research. The United States Military Academy at West Point, for instance, is modeled after it. The Ecole Polytechnique was a model for the modern university, in classroom instruction and examination, and in having active researchers as instructors. Teachers developed textbooks out of course lectures; such texts are the immediate ancestors of our modern college textbooks. Unfortunately, Germain could not avail herself of this splendid new institution, which barred women from attending, but she did obtain lecture notes. Using the name of an acquaintance registered as a student at the Ecole Polytechnique, Antoine-August Le Blanc, Germain submitted a report on analysis to Lagrange. Lagrange was impressed with the originality and insight of "M. Le Blanc," and wished to meet him. Discovering the true identity of the report's author did not diminish Lagrange's opinion of the work; he provided Germain with support and encouragement for many years.

Germain continued her studies in mathematics, and began to make original contributions to research. She is best known for her work on the theory of elasticity, in particular the theory of vibrating surfaces, for which she was awarded a prize of the Academy of Sciences in 1816. An excellent account of Sophie Germain's life and work is [23]. Shorter accounts can be found in [35, 42, 79].

Despite her accomplishments in elasticity theory, her major work was in number theory. In fact, it was number theory that was her true love, which occupied her throughout her life. Early on, she began studying Legendre's *Theory of Numbers*, published in 1789, and wrote to him about it, thereby initiating an extensive correspondence. Germain's interest in number theory coincided with the 1801 publication of Gauss's *Disquisitiones Arithmeticae*, the groundbreaking work that propelled number theory into the very center of nineteenth-century mathematics. After carefully studying the *Disquisitiones*, Germain initiated a correspondence with Gauss, again using the name Le Blanc, praising the *Arithmeticae* and enclosing some of her own results for his evaluation. Gauss was impressed with her efforts, going so far as to mention Monsieur LeBlanc very favorably to other scientists.

When, in 1807, Germain revealed her true identity to Gauss, he responded:

But how can I describe my astonishment and admiration on seeing my esteemed correspondent Monsieur LeBlanc metamorphosed into this celebrated person, yielding a copy so brilliant it is hard to believe? The taste for the abstract sciences in general and, above all, for the mysteries of numbers, is very rare: this is not surprising, since the charms of this sublime science in all their beauty reveal themselves only to those who have the courage to fathom them. But when a woman, because of her sex, our customs and prejudices, encounters infinitely more obstacles than men, in familiarizing herself with their knotty problems, yet overcomes these fetters and penetrates that which is most hidden, she doubtless has the most noble courage, extraordinary talent, and superior genius. Nothing could prove to me in a more flattering and less equivocal way that the attractions of that science, which have added so much joy to my life, are not chimerical, than the favor with which you have honored it.

The scientific notes with which your letters are so richly filled have given me a thousand pleasures. I have studied them with attention and I admire the ease with which you penetrate all branches of arithmetic, and the wisdom with which you generalize and perfect [23, p. 25].

Germain outlined her strategy for a general proof of FLT in a long letter to Gauss, written on May 12, 1819, after a ten-year hiatus in their correspondence. Before delving into an explanation of her work she expresses her long-term devotion to the study of number theory:

Although I have worked for some time on the theory of vibrating surfaces. . .I have never ceased thinking about the theory of numbers. I will give you a sense of my absorption with this area of research by admitting to you that even without any hope of success, I still prefer it to other work which might interest me while I think about it, and is sure to yield results.

Long before our Academy proposed a prize for a proof of the impossibility of the Fermat equation, this type of challenge, which was brought to modern theories by a geometer who was deprived of the resources we possess today, tormented me often. I have a vague inkling of a connection between the theory of residues and the famous equation; I believe I spoke to you of this idea a long time ago, because it struck me as soon as I read your book.

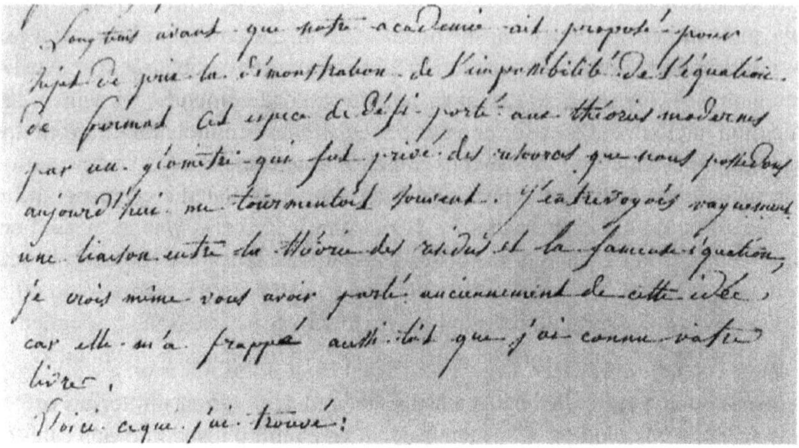

Longtemps avant que notre Académie ait proposé pour sujet de prix la démonstration de l'impossibilité de l'équation de Fermat cet espèce de défi porté aux théories modernes par un géomètre qui fut privé des ressources que nous possédons aujourd'hui me tourmentait souvent. J'entrevoyais vaguement une liaison entre la théorie des résidus et la fameuse équation, je crois même vous avoir parlé anciennement de cette idée, car elle m'a frappée aussitôt que j'ai connu votre livre.

Voici ce que j'ai trouvé:

PHOTO 4.12. Germain's 1819 letter to Gauss.

Here is what I have found [69]:

She then goes on to make a simple but very important observation that is central to her method. Let p be an odd prime.

Basic Lemma:

If the Fermat equation for exponent p has a solution, and if θ is a prime number with no nonzero consecutive pth power residues modulo θ, then θ necessarily divides one of the numbers x, y, or z.

To see why this is true, first note that what is meant by a pth power residue modulo θ is simply the remainder, modulo θ, of a pth power. (See the Appendix for a brief introduction to congruence arithmetic.) So now suppose the Fermat equation $x^p + y^p = z^p$ has a solution, and suppose that none of x, y, or z is divisible by θ. Thus, modulo θ, we can divide by any of x, y, z (see Proposition 4 of the Appendix). Letting a be a multiplicative inverse for x modulo θ we obtain the congruence

$$(ax)^p + (ay)^p \equiv (az)^p \pmod{\theta},$$

and thus $1 + (ay)^p \equiv (az)^p$. Thus the residues of $(ay)^p$ and $(az)^p$ will be consecutive. Notice that they are also nonzero, since θ does not divide a, y, or z. This contradicts the assumption on θ, proving the assertion.

Germain then concludes that if for a fixed p one could find infinitely many primes θ satisfying the condition that θ has no nonzero consecutive pth power residues modulo θ, then, by the previous observation, each of these would have to divide one of x, y, z, and thus one of these three numbers would be divisible by infinitely many primes, which is absurd. This would prove FLT for that exponent.

Despite much effort, she never succeeded in proving FLT by this approach for even a single exponent. However, she invented a method for producing many primes θ satisfying the above condition. For any particular exponent, her method may then show that any solutions to the Fermat equation would have to be quite large. She made many such applications of her method in her manuscripts [70]. For instance, for $p = 5$, she showed that any solutions to the Fermat equation would have to be at least 30 decimal digits in size! As she says in her letter to Gauss, "You can easily imagine, Monsieur, that I must have been able to prove that this equation is only possible for numbers whose size frightens the imagination. . . .But all this is still nothing; it requires the infinite and not just the very large."

Sophie Germain never published her work on FLT. One might speculate that her experience with the establishment at the Paris Academy of Sciences played an important role in this decision. During her time, it was common to publish papers as memoirs of the Academy. After several disputes relating to the publication of her prize-winning work on elasticity theory, she ended up publishing that work at her own expense. A renewed battle over the publication of her number-theoretic work might have been too unpleasant for her to contemplate. It should be noted that a few years before her death she did publish a short paper on number theory in the recently founded *Journal für die reine und angewandte Mathematik*, a private German scientific journal.

The only commonly known result of Germain's appeared in 1825, as part of a supplement to the second edition of Legendre's *Theory of Numbers*. The supplement also appeared as a *Memoir* of the Royal Academy of Sciences of the Institut de France in 1827 [107]. In this supplement, Legendre presents his own proof (the first) for the $p = 5$ case of FLT, along with part of Germain's work, explicitly credited to her in a footnote. The reference in this footnote is commonly considered to be her only contribution to FLT and is known as "Sophie Germain's Theorem." In fact, her work is considerably more extensive than this result [102].

In order to state her theorem, first observe that it is enough to consider Fermat's equation for odd prime exponents (see Exercise 4.10 of the Introduction). Recall from the Introduction that we may furthermore assume that x, y, and z are relatively prime. Germain's theorem addresses the existence of solutions in which none of the three numbers x, y, z is divisible by the prime exponent p. Today this is called Case I of FLT. We state her result in modern terminology.

Sophie Germain's Theorem: *If p is an odd prime, and if there exists an auxiliary prime θ with the properties that*

1. *p is not a pth power modulo θ, and*
2. *the equation $r' \equiv r + 1$ modulo θ cannot be satisfied for any two pth power residues,*

then Case I of Fermat's Last Theorem is true for p.

Note that Condition 2 is identical to the condition she mentions in her letter to Gauss, that there should not exist any nonzero consecutive pth power residues modulo θ.

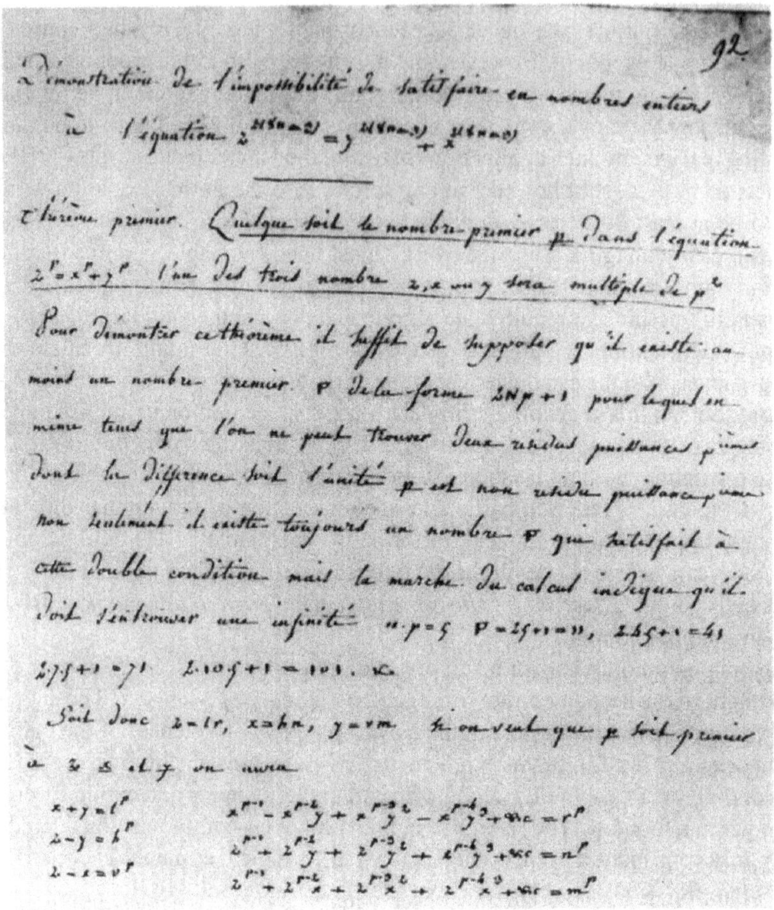
92

Démonstration de l'impossibilité de satisfaire en nombres entiers à l'équation $z^{2(8n\pm3)} = y^{2(8n\pm3)} + x^{2(8n\pm3)}$

Théorème premier. Quelque soit le nombre premier p dans l'équation $z^p = x^p + y^p$ l'un des trois nombres z, x ou y sera multiple de p^2

Pour démontrer ce théorème il suffit de supposer qu'il existe au moins un nombre premier P de la forme $2Np+1$ pour lequel en même temps que l'on ne peut trouver deux résidus puissances $p^{ièmes}$ dont la différence soit l'unité p est non résidu puissance $p^{ième}$ non seulement il existe toujours un nombre P qui satisfait à cette double condition mais la marche du calcul indique qu'il doit s'en trouver une infinité $n.p=5$ $P=2.5+1=11$, $2.4.5+1=41$ $2.7.5+1=71$ $2.10.5+1=101$. &c

Soit donc $z=lr$, $x=hn$, $y=vm$ si on veut que p soit premier à z x et y on aura

$x+y=l^p$ $x^{p-1}-x^{p-2}y+x^{p-3}y^2-x^{p-4}y^3+\&c=r^p$

$z-y=h^p$ $z^{p-1}+z^{p-2}y+z^{p-3}y^2+z^{p-4}y^3+\&c=n^p$

$z-x=v^p$ $z^{p-1}+z^{p-2}x+z^{p-3}x^2+z^{p-4}x^3+\&c=m^p$

PHOTO 4.13. Germain's manuscript.

Germain also developed methods to verify the hypotheses of her theorem and applied them to do so for all primes less than 100.

Her result is implicit in a theorem from a handwritten manuscript in the *Manuscrits de Sophie Germain*, MS. FR 9114, in the Bibliothèque Nationale in Paris, beginning on page 92.

Germain, from a manuscript entitled *Demonstration of the impossibility to satisfy in integers the equation $z^{2(8n\pm3)} = y^{2(8n\pm3)} + x^{2(8n\pm3)}$*

First Theorem. *For any [odd] prime number p in the equation $z^p = x^p + y^p$ one of the three numbers z, x, or y will be a multiple of p^2.*

To prove this theorem it suffices to suppose that there exists at least one prime number θ of the form $2Np + 1$ for which at the same time one cannot find two pth power residues whose difference is one, and p is not a pth power residue. Not only does there always exist a number θ satisfying these two conditions, but the course of calculation indicates that there must be an infinite number of them. For example, if $p = 5$, then $\theta = 2 \cdot 5 + 1 = 11, 2 \cdot 4 \cdot 5 + 1 = 41, 2 \cdot 7 \cdot 5 + 1 = 71$, $2 \cdot 10 \cdot 5 + 1 = 101$, etc.

Let therefore $z = lr$, $x = hn$, $y = vm$. If one assumes that p is [relatively] prime to z, x, and y, then one will have that

$$x + y = l^p \qquad x^{p-1} - x^{p-2}y + x^{p-3}y^2 - x^{p-4}y^3 + \text{ etc } \quad = r^p,$$
$$z - y = h^p \qquad z^{p-1} + z^{p-2}y + z^{p-3}y^2 + z^{p-4}y^3 + \text{ etc } \quad = n^p,$$
$$z - x = v^p \qquad z^{p-1} + z^{p-2}x + z^{p-3}x^2 + z^{p-4}x^3 + \text{ etc } \quad = m^p.$$

Since we have assumed that there are no two pth power residues modulo θ whose difference is one, it follows that in the equation $z^p = x^p + y^p$ one of the numbers x, y, z is necessarily a multiple of θ. To make a choice, let us take $z \equiv 0 \pmod{\theta}$. Thus one has $l^p + h^p + v^p \equiv 0 \pmod{\theta}$. It is therefore necessary, again, that one of the numbers l, h, v be a multiple of θ; because $z = lr \equiv 0$ it can only be l. And as a result $x \equiv -v^p$, $y \equiv -h^p$, $x + y \equiv 0 \pmod{\theta}$. Consequently $px^{p-1} \equiv pv^{p(p-1)} \equiv r^p$. That is to say, p is a pth power residue, contrary to the hypothesis.

The first sentence after the statement of the theorem must actually be considered a hypothesis. While Germain believed that an auxiliary prime θ satisfying the assumed conditions exists for any prime number p, she never succeeded in showing this. Also, she assumes that x, y, and z are relatively prime, since any solution of the Fermat equation in which x, y, and z are not relatively prime can be used to find a solution x', y', z' such that x', y', and z' are relatively prime.

Germain's reasoning is very terse and requires substantial work on the part of the reader. In order to follow her argument in the second paragraph, let us first derive the displayed equations, which are to follow from her additional assumption that p does not divide x, y, or z. From this assumption, she will derive a contradiction; i.e., she will show that it is impossible for p not to divide one of x, y, or z. We note that in the portion of the manuscript quoted above, Germain shows only that one of x, y, or z must be divisible by p rather than p^2 as in the statement of the theorem. Begin by factoring:

$$x^p + y^p = (x + y)\varphi(x, y),$$

where

$$\varphi(x, y) = x^{p-1} - x^{p-2}y + x^{p-3}y^2 - x^{p-4}y^3 + \cdots + y^{p-1}.$$

Observe that this factorization holds because p is odd. We next show that $x + y$ and $\varphi(x, y)$ are relatively prime. If not, let q be a prime dividing both. Then $y = -x +$ some multiple of q, and substitution yields $\varphi(x, y) = px^{p-1} +$ a multiple of q. From this we see that either p or x must be divisible by q. If p is divisible by q, then it equals q, since both are prime. Thus $p = q$ divides $x + y$, hence also z.

But this contradicts Germain's assumption that p is prime to x, y, and z. If, on the other hand, x were divisible by q, then x and $x + y$ would both have q as common factor, contradicting the assumption that x and y are relatively prime.

Now, since the product of the relatively prime numbers $x + y$ and $\varphi(x, y)$ is the pth power z^p, it must be that each of them is itself a pth power (see Appendix). Thus, we may write them as l^p and r^p respectively, as shown in the displayed equations. Multiplying these two together yields $z^p = (x + y)\varphi(x, y) = l^p r^p = (lr)^p$, and so $z = lr$, as Germain claims at the beginning of her argument. The other equations follow similarly from $z^p - y^p = x^p$ and $z^p - x^p = y^p$.

Her next assertion, that one of x, y, z is a multiple of θ, follows from the Basic Lemma. The argument continues by assuming that it is z. The reader may check that if in fact it were x or y, everything that follows could be carried out in a similar manner. Adding the left displayed equations, we obtain

$$l^p + h^p + v^p = 2z \equiv 0 \pmod{\theta}.$$

Imitating the proof of the Basic Lemma, the hypothesis on θ insures that one of l, h, or v is a multiple of θ. Her next assertion implicitly assumes that we are dealing only with primitive solutions to the Fermat equation, that is, x, y, and z are relatively prime. If either h or v were divisible by θ, then y or x would have the factor θ in common with z, violating primitivity.

From the congruence $x + y = l^p \equiv 0 \pmod{\theta}$ she obtains, by substituting $y \equiv -x$ in the top right equation, that

$$r^p \equiv px^{p-1} \equiv p(-v^p)^{p-1} = pv^{p(p-1)}.$$

Since v is not divisible by θ we may divide by $v^{p(p-1)}$ (see Appendix), obtaining

$$p \equiv \left(\frac{r}{v^{p-1}}\right)^p.$$

As Germain observes, this contradicts one of the hypotheses of the theorem.

Notice again that at this point Germain has only proven that one of the numbers x, y, z is divisible by p rather than p^2. It is this weaker version of the theorem that is called Sophie Germain's Theorem in the literature [47, p. 64]. Moreover, this result is responsible for the division of possible solutions into two types: Case I solutions are those where x, y, z are all prime to the exponent p, whereas Case II solutions are those where p divides one of x, y, z. Thus, Germain's Theorem proves the nonexistence of Case I solutions whenever an auxiliary prime θ can be found satisfying the required conditions, and she succeeded in doing this up to exponent 97.

Much of Germain's other work on Fermat's Last Theorem was directed toward developing methods for finding such auxiliary primes θ. For instance, she developed a method for proving that for any odd prime p, if $\theta = 2Np + 1$ is also prime and N is not a multiple of 3, then it satisfies the condition that there are no consecutive pth power residues modulo θ. She made a detailed study for $N \leq 10$, however without verifying that her method worked in general [70, pp. 198 ff.],[102].

An extension of her method was used as late as 1985 to prove that Case I holds for infinitely many prime exponents [2], testifying to the importance and depth of her ideas. Germain's ideas have not lost their relevance after almost two hundred years.

Exercise 4.26: Read about Germain's work on elasticity theory and why she became interested in it.

Exercise 4.27: Read about the work of other female mathematicians before and after Germain.

Exercise 4.28: Find all quadratic residues modulo 3. Find all 7th power residues modulo 29.

Exercise 4.29: Use Germain's Theorem to show that there are no Case I solutions to the Fermat equation for exponent 3 by showing that 3 is not a cube modulo 7, and that no two nonzero third power residues differ by 1.

Exercise 4.30: Use Germain's Theorem to show that there are no Case I solutions to the Fermat equation for exponent 7.

Exercise 4.31: Can a prime θ of the form $2Np + 1$ satisfy Germain's nonconsecutivity condition if N is a multiple of 3?

Exercise 4.32: Verify the hypothesis of Germain's Theorem for the first prime larger than 100.

4.5 Kummer and the Dawn of Algebraic Number Theory

Germain's general approach notwithstanding, progress on the problem continued only very slowly, with proofs for specific exponents. Credit for exponent 5 was shared between Legendre and Dirichlet. By the time Dirichlet succeeded in producing a proof for exponent 14 in 1832, and the French mathematician Gabriel Lamé for exponent 7 in 1839, it had become clear that their methods seemed unlikely to lead further. A whole new approach was needed, based on different principles.

Such a new method was proposed by Lamé, who initiated a flurry of exchanges in the proceedings of the Paris Academy of Sciences by announcing at the March 1 meeting in 1847 a proof of Fermat's Last Theorem [47, 4.1]. Lamé observed that one difficulty encountered in trying to generalize previous approaches to the problem was that one of the factors in the decomposition

$$x^n + y^n = (x + y)(x^{n-1} - x^{n-2}y + x^{n-3}y^2 + \cdots + y^{n-1})$$

has increasingly high degree. Here, and subsequently, n is always odd. He noted that this difficulty can be overcome by factoring the right-hand side completely into linear factors, using certain complex numbers.

This can be accomplished with the help of the complex number $r = e^{2\pi i/n}$. This number is called a *primitive nth root of unity*, since $r^n = 1$ and no smaller power of r has this property. We obtain the factorization

$$x^n + y^n = (x + y)(x + ry)(x + r^2 y) \cdots (x + r^{n-1} y),$$

as follows. First note that $X^n - 1 = (X - 1)(X - r)(X - r^2) \cdots (X - r^{n-1})$ because both sides are nth-degree polynomials with the same roots. If we set $X = -x/y$ and multiply by $-y^n$, then the linear factorization of $x^n + y^n$ follows. Lamé now planned to use the same techniques we have already seen in Euler's and Germain's work to push through a general proof by infinite descent. In particular, if x and y were solutions to the Fermat equation for which the linear factors above are relatively prime, he deduced that since their product is equal to the nth power z^n, each linear factor is therefore an nth power itself, leading to an infinite descent argument. Such an argument assumes that if a product of relatively prime numbers is an nth power, then each of the factors is also an nth power. The proof of this fact requires unique factorization into primes, which is assured for integers by the Fundamental Theorem of Arithmetic (see Appendix, Lemma 4.3). Lamé was confident that this theorem held true for the more general class of complex numbers that he had employed.

After several weeks of claims and counterclaims in the Paris Academy, on May 24 a letter from the German mathematician Ernst Kummer [100] brought this exchange to an abrupt end.

Ernst Kummer (1810–1893) was one of the foremost number theorists of the nineteenth century. He received his doctorate in mathematics from the University of Halle in 1828. After teaching high school for ten years, he was appointed full professor at the University of Breslau, in present-day Poland, in 1842. When Dirichlet left Berlin in 1855 to succeed Gauss in Göttingen, Kummer was appointed as his successor. His very distinguished career as researcher, teacher, and administrator came to an abrupt end in 1882, when he announced his retirement, giving a weakening of his memory as the reason. Besides number theory, Kummer also made very significant contributions to function theory and geometry.

Ernst Kummer, from

A Letter to Joseph Liouville,
Journal de Mathématiques Pures et Appliquées XII, 136 (1847)

Breslau, April 28, 1847

At the suggestion of my friend Monsieur Lejeune-Dirichlet, I take the liberty of sending you several copies of an essay I wrote three years ago, on the occasion of the centenary of the University of Koenigsberg, and of another essay by one of my friends and disciples, Monsieur Kronecker, a young distinuished geometer. In these essays, which I ask you to accept as a sign of my greatest admiration, you will find some developments on several points in the theory of complex numbers

PHOTO 4.14. Kummer.

composed of roots of unity; that is, of the equation $r^n = 1$, which has recently been the subject of several discussions in your illustrious academy, occasioned by the attempted proof of the theorem of Fermat, proposed by Monsieur Lamé. Regarding the elementary proposition for complex numbers, *that a composite complex number can only be decomposed into prime factors in one way*, which you justifiably criticize in this defective proof, along with several other points, I can assure you, *that it does not hold in general* when one considers complex numbers of the form $\alpha_0 + \alpha_1 r + \alpha_2 r^2 + \cdots + \alpha_{n-1} r^{n-1}$, but one can save it by introducing complex numbers of a new kind, which I call an *ideal complex number*. The results of my research on this subject were communicated to the Berlin Academy and published in the Proceedings (March 1846); a memoir on the same subject will appear shortly in Crelle's Journal. The applications of this theory to the proof of Fermat's Theorem have occupied me for a long time, and I have succeeded in making the impossibility of the equation $x^n - y^n = z^n$ a consequence of two properties of the prime number n. Thus, all that is left to do is to study whether they are satisfied by all prime numbers. In case you deem these results of interest you may find them in this month's Proceedings of the Berlin Academy of Sciences.

For small values of n Lamé indeed was right; unique factorization does hold. For $n \geq 23$, however, it fails, as Kummer had claimed. A very detailed and engaging account of these developments can be found in [47].

Kummer began a program to understand the failure of unique factorization, and ended up proposing a whole new way of looking at this phenomenon. He concluded that the problem simply lay in the fact that the concept of number was too narrow. He boldly constructed a brave new world of "ideal numbers" in which unique factorization was restored. The trade-off was that ideal numbers required a new level of abstraction. In order to explain Kummer's theory, we first need to understand what it means for unique factorization to fail in a number system. The number systems of relevance to FLT are those involving roots of unity, but the failure of unique factorization appears in many other number systems as well. We first consider a number system involving $\sqrt{-5}$. We will use $\mathbf{Z}$ to denote the integers. In $\mathbf{Z}$, the Fundamental Theorem of Arithmetic asserts that unique factorization holds: that is, there is exactly one way to write an integer $n > 1$ as a product of prime numbers (up to possible rearrangement of the prime factors). But what exactly is a prime number? A common definition is that a number p (in $\mathbf{Z}$) is a prime if

$$m \mid p \text{ implies that } m = \pm 1 \text{ or } m = \pm p.$$

Though this definition is sufficient in $\mathbf{Z}$, it isn't good enough in other number systems, as we shall see. We will call a number p that satisfies this condition *irreducible*, and reserve the term *prime* for p ($\neq 0, \pm 1$) such that

$$p \mid mn \text{ implies that either } p \mid m \text{ or } p \mid n.$$

In $\mathbf{Z}$, the two conditions are equivalent (Exercise 4.33).

Now consider the number system $\mathbf{Z}[\sqrt{-5}]$, that is, the set of complex numbers of the form $\{a + b\sqrt{-5} \mid a, b \in \mathbf{Z}\}$. This system is closed under addition and multiplication of complex numbers; that is, the sum and product of any two numbers of this form is again a number of this form (Exercise 4.34). Thus, the set $\mathbf{Z}[\sqrt{-5}]$ gives us an example of a number system that is larger than $\mathbf{Z}$. Define the *norm function* N: $\mathbf{Z}[\sqrt{-5}] \rightarrow \mathbf{Z}$ by $N(a + b\sqrt{-5}) = a^2 + 5b^2$. This function, which assigns an integer to every number in $\mathbf{Z}[\sqrt{-5}]$, has very special properties: it is multiplicative, that is, for $x, y \in \mathbf{Z}[\sqrt{-5}]$, we have $N(xy) = N(x)N(y)$; and if $x \in \mathbf{Z}[\sqrt{-5}]$ and $N(x)$ is a prime integer, then x is irreducible (Exercise 4.35). The norm function allows us to draw conclusions about numbers in $\mathbf{Z}[\sqrt{-5}]$ from properties of their norm in $\mathbf{Z}$.

Now observe that in $\mathbf{Z}[\sqrt{-5}]$ we have two different factorizations of the integer 6:

$$2 \cdot 3 = 6 = (1 + \sqrt{-5})(1 - \sqrt{-5}).$$

Furthermore, all four factors are irreducible but not prime in $\mathbf{Z}[\sqrt{-5}]$ (Exercise 4.36). It is in fact not always possible to write a number in $\mathbf{Z}[\sqrt{-5}]$ as a product of primes (Exercise 4.37).

Closer to our topic, consider the number system of complex numbers built from a primitive 23rd root of unity $\omega_{23} = \omega$, that is, $\omega^{23} = 1$ and no smaller power of ω is equal to 1. Then ω satisfies the equation $x^{23} - 1 = 0$. After factoring the left side as

$$(x - 1)(x^{22} + x^{21} + \cdots + x^2 + x + 1),$$

we see that ω is a root of the second factor. Then consider the set $\mathbf{Z}[\omega]$ consisting of all complex numbers of the form

$$1 + a_1\omega + a_2\omega^2 + \cdots + a_{21}\omega^{21},$$

with the coefficients $a_i \in \mathbf{Z}$. (Note that ω is a root of the polynomial $x^{22}+\cdots+x+1$, so that ω^{22} can be expressed in terms of the smaller powers of ω. This is the reason why it is not needed to generate the numbers in $\mathbf{Z}[\omega]$.) It is easy to see that this system is closed under addition and multiplication of complex numbers. As Kummer pointed out to Liouville, unique factorization holds in all number systems $\mathbf{Z}[\omega_n]$, for $n \leq 19$, but fails for $n = 23$ (and all larger primes n). As an example consider the product

$$(*)\quad (1 + \omega^2 + \omega^4 + \omega^5 + \omega^6 + \omega^{10} + \omega^{11})(1 + \omega + \omega^5 + \omega^6 + \omega^7 + \omega^9 + \omega^{11})$$

in $\mathbf{Z}[\omega_{23}]$. While the product is divisible by 2 (Exercise 4.38), neither factor is. One can furthermore show that 2 is irreducible in $\mathbf{Z}[\omega_{23}]$. This shows that unique factorization fails here too. (For details see [115, p. 11].)

In order to introduce Kummer's idea of an "ideal number," let us consider the following artificial but illustrative example: Suppose we know how to multiply only positive integers that are congruent to 1 modulo 5. Note that when we multiply any two numbers in this set, we get another in the set; so we can talk about multiplication and factorization inside this subworld of the integers. We can also discuss irreducible and prime elements, as, for instance, 6 is irreducible, since its only divisors are 1 and itself. Further, 6 is not prime, since $6 \cdot 56 = 336 = 16 \cdot 21$, but neither 16 nor 21 is divisible by 6. Also, 56, 16, and 21 are all irreducible in this system; so factorization is not unique. This failure of unique factorization stems from the fact that in the larger world of all the integers, 6 is *not* irreducible, since it is equal to $2 \cdot 3$ (note that 2 and 3 are prime). In our subworld, the "real" prime factors 2 and 3 were hidden from us.

Kummer's fix for the problem of nonuniqueness of factorization is based on this idea of the possible existence of "hidden" divisors. For example, in Exercise 4.36 we see that

$$6 = 2 \cdot 3 = (1 + \sqrt{-5})(1 - \sqrt{-5}).$$

We might postulate that 2 and $1 + \sqrt{-5}$, say, have some common divisor. This common divisor would be what Kummer called an "ideal number."

Let us return to "ordinary" numbers for a moment and consider the greatest common divisor of 6 and 9, which is 3. Then 3 will divide any integer of the form $6m + 9n$, for m, n arbitrary integers. Conversely, this set of linear combinations of 6 and 9 determines their greatest common divisor uniquely (Exercise 4.39). The same is true for ideal numbers. It turns out that one way to represent these ideal common divisors is as sets of elements that have the ideal number as divisor. For instance, denote the common divisor of 2 and $1 + \sqrt{-5}$ by A. Then one can represent A by the set

$$A = \{2m + (1 + \sqrt{-5})n \mid m, n \in \mathbf{Z}[\sqrt{-5}]\}.$$

That is, A is the set of all linear combinations of 2 and $1 + \sqrt{-5}$ with coefficients in $\mathbf{Z}[\sqrt{-5}]$. Let us write $A = \langle 2, 1 + \sqrt{-5} \rangle$. One can show that A is closed under addition and under multiplication by arbitrary elements from $\mathbf{Z}[\sqrt{-5}]$ (Exercise 4.40). This view of ideal numbers was introduced by Dedekind, and has become a central concept in modern number theory. As an example of how to calculate with ideal numbers, we compute the square of the ideal number A. Thinking of $A \cdot A$ as the set of elements of $\mathbf{Z}[\sqrt{-5}]$ that have this ideal number as divisor, we get that

$$A \cdot A = \langle 4, 2(1 + \sqrt{-5}), -4 + 2\sqrt{-5} \rangle.$$

It is clear that every element in $A \cdot A$ is divisible by 2. But $A \cdot A$ actually contains 2, because it contains 4 and therefore $2\sqrt{-5}$. Subtracting from $2(1 + \sqrt{-5})$ we obtain 2. Thus we have shown that

$$A \cdot A = \langle 2 \rangle.$$

Here we view $\langle 2 \rangle$ as the ideal number of all elements of $\mathbf{Z}[\sqrt{-5}]$ that are divisible by 2.

Now let B, C denote common divisors of 3 and $1 + \sqrt{-5}$, and of 3 and $1 - \sqrt{-5}$, respectively. Then $A \cdot B = \langle 1 + \sqrt{-5} \rangle$, $B \cdot C = \langle 3 \rangle$, and $A \cdot C = \langle 1 - \sqrt{-5} \rangle$; so

$$(6) = \langle 2 \rangle \langle 3 \rangle = A^2 BC = ABAC = \langle 1 + \sqrt{-5} \rangle \langle 1 - \sqrt{-5} \rangle$$

(Exercise 4.41). Thus, the apparent lack of unique factorization for numbers in $\mathbf{Z}[\sqrt{-5}]$ turns out to be simply a matter of grouping the factors of a product differently, when viewed in the much finer world of ideal numbers. With this stroke of genius Kummer forever transformed number theory and founded what is now called algebraic number theory. Subsequently, the idea behind ideal numbers was generalized to other algebraic systems and is now at the heart of many constructions in abstract algebra.

Once he had his new theory in place, Kummer returned to the Fermat equation, and with ingenious arguments managed to prove Fermat's Last Theorem for all prime exponents that are *regular*. It would lead us too far afield to explain here the condition that makes a prime regular and its significance. But Kummer showed that all primes less than 100, except 37, 59, and 67, are regular. Extending his argument to deal with those three nonregular primes, he proved that FLT holds for all prime exponents less than 100. A detailed description of Kummer's work can found in [163], for instance. It is still an open problem whether there are infinitely many regular primes. However, we do know that there are infinitely many primes that are not regular.

Exercise 4.33: Show that the conditions for irreducibility and primality are equivalent in $\mathbf{Z}$.

Exercise 4.34: Denote by $\mathbf{Z}[\sqrt{-5}]$ the set of complex numbers of the form $\{a + b\sqrt{-5} \mid a, b \in \mathbf{Z}\}$. Show that $\mathbf{Z}[\sqrt{-5}]$ is closed under addition and multiplication.

Exercise 4.35:

1. Show that N is multiplicative; i.e., for $x, y \in \mathbf{Z}[\sqrt{-5}]$, we have $N(xy) = N(x)N(y)$.
2. Show that if $x \in \mathbf{Z}[\sqrt{-5}]$ and $N(x)$ is a prime integer, then x is irreducible.

Exercise 4.36: Use the norm function and the fact that $6 = 2 \cdot 3 = (1 + \sqrt{-5})(1 - \sqrt{-5})$ to show that $2, 3, 1 + \sqrt{-5}, 1 - \sqrt{-5}$ are irreducible but not prime in $\mathbf{Z}[\sqrt{-5}]$.

Exercise 4.37: Show that in $\mathbf{Z}[\sqrt{-5}]$ it is not always possible to write a number as a product of primes.

Exercise 4.38: Show that the product $(*)$ in $\mathbf{Z}[\omega_{23}]$ is divisible by 2.

Exercise 4.39: Let m, n be integers. Show that the greatest common divisor of m and n is the only positive integer that divides all numbers of the form $xm + yn$, with x, y arbitrary integers.

Exercise 4.40: Show that the set $A = \{2m + (1 + \sqrt{-5})n \mid m, n \in \mathbf{Z}[\sqrt{-5}]\}$ is closed under addition and under multiplication by arbitrary elements from $\mathbf{Z}[\sqrt{-5}]$.

Exercise 4.41: Let B, C denote common divisors of 3 and $1 + \sqrt{-5}$, and of 3 and $1 - \sqrt{-5}$, respectively. Show that $AB = \langle 1 + \sqrt{-5} \rangle$, $BC = \langle 3 \rangle$, and $AC = \langle 1 - \sqrt{-5} \rangle$; so

$$\langle 6 \rangle = \langle 2 \rangle \langle 3 \rangle = A^2 BC = ABAC = \langle 1 + \sqrt{-5} \rangle \langle 1 - \sqrt{-5} \rangle.$$

4.6   Appendix on Congruences

This appendix contains a short introduction to congruence arithmetic and some basic number-theoretic results. For a more extensive treatment the reader is encouraged to consult, for instance, [6].

For two integers a and b and a positive integer n, we say that a is *congruent to b modulo n*, denoted by

$$a \equiv b \bmod n,$$

if n divides $a - b$, or, equivalently, if there is an integer k such that $a = kn + b$. We will use the notation $x \mid y$ to indicate that x divides y; that is, there is an integer m such that $mx = y$. The greatest common divisor of m and n will be denoted gcd(m, n). The notion of congruence was first treated systematically by Gauss in his *Disquisitiones Arithmeticae* [68], although it can already be found in the work of Euler, Lagrange, and Legendre. The notation above is due to Gauss, however [20, p. 550], [97, pp. 813–14].

Example. Since $4 \mid 7 - 15$, we have that $7 \equiv 15$ (mod 4). Note that since $2 \mid 4$, we also know that $7 \equiv 15$ mod 2. However, $7 \not\equiv 15$ mod 5, since $5 \nmid 7 - 15$.

Example. An everyday example of congruence is found in the computation of time. For instance (using a 12-hour clock), if it is 10 o'clock now, we know in 5 hours it will be 3 o'clock. What we have really done is compute $10 + 5 = 15 \equiv 3$ mod 12.

Proposition 1: Let n be a positive integer, and let a, b, c be integers. Then

1. $a \equiv a$ mod n.
2. If $a \equiv b$ mod n, then $b \equiv a$ mod n.
3. If $a \equiv b$ mod n and $b \equiv c$ mod n, then $a \equiv c$ mod n.

Proof. Exercise 4.43.

(The three conditions of Proposition 1 show that congruence is an *equivalence relation*. See Exercise 4.44.) For any integer a, the *equivalence* (or *residue*) *class* of a modulo n consists of all integers that are congruent to a modulo n. Using the division algorithm, we can find integers q, r such that

$$a = qn + r \text{ where } 0 \leq r < n.$$

So a is in the residue class of some nonnegative integer r less than n; and since q and r are uniquely determined by the algorithm, a is in exactly one such class. The integer r is called the *residue* of a.

Example. The residue of 7 modulo 4 is 3, since $7 = 1(4) + 3$, and the residue of 15 modulo 4 is also 3, since $15 = 3(4) + 3$. So $7 \equiv 3$ mod 4 and $15 \equiv 3$ mod 4; Proposition 1 implies that $7 \equiv 15$ mod 4, as we found in Example 1. The residue of 7 modulo 5 is 2, since $7 = 1(5) + 2$; and the residue of 15 modulo 5 is 0, since $15 = 3(5) + 0$. So $7 \equiv 2$ mod 5 and $15 \equiv 0$ mod 5, and since clearly $2 \not\equiv 0$ mod 5, we again discover that $7 \not\equiv 15$ mod 5.

Proposition 2: Let n be a positive integer, and let a, b, c, d be integers. Then

1. If $a \equiv b$ mod n and $c \equiv d$ mod n, then $a + c \equiv b + d$ mod n.
2. If $a \equiv b$ mod n and $c \equiv d$ mod n, then $ac \equiv bd$ mod n.

Proof. For part 1, we know that $n \mid a - b$ and $n \mid c - d$; so $n \mid (a - b) + (c - d) = (a + c) - (b + d)$. That is, $a + c \equiv b + d$ mod n.

In part 2, since $n \mid a - b$ and $n \mid c - d$, we have $n \mid (a - b)c + (c - d)b = ac - bd$; so $ac \equiv bd$ mod n.

Example. Proposition 2 essentially says that addition and multiplication make sense modulo n. For instance, suppose we are working modulo 4, and we wish to add 1247 and 10118. Suppose we add first: $1247 + 10118 = 11365$, and then find the residue of the result modulo 4: $11365 = 2841(4) + 1$. So $1247 + 10118 \equiv 1$ mod 4. If we instead find the residues of 1247 and 10118 first, and then add, we will get the same result: $1247 \equiv 3$ mod 4 and $10118 \equiv 2$ mod 4. So $1247 + 10118 \equiv$

$3 + 2 \equiv 5 \equiv 1 \bmod 4$. Multiplication works similarly: $1247(10118) = 12617146 = (3154286)4 + 2$; so $1247(10118) \equiv 2 \bmod 4$. We also have $1245(10118) \equiv 3(2) \equiv 6 \equiv 2 \bmod 4$. In many cases, this property of multiplication modulo n can greatly simplify computation. Suppose we wish to find $56^{99} \bmod 5$. Clearly, we would prefer not to have to compute 56^{99}! We get around it by observing that $56 \equiv 1 \bmod 5$; so $56^{99} \equiv 1^{99} \equiv 1 \bmod 5$. To compute $2^{99} \bmod 7$, note that $2^{99} = (2^3)^{33} = 8^{33}$; and as $8 \equiv 1 \bmod 7$, we get $2^{99} \equiv 1^{33} \equiv 1 \bmod 7$.

Part 2 of Proposition 2 implies that if we multiply both sides of a congruence by the same number, the congruence is preserved. Unfortunately, division is not as nice, as the next example shows.

Example. Note that $9 \equiv 15 \bmod 6$; i.e., $3 \cdot 3 \equiv 3 \cdot 5 \bmod 6$. If we attempt to "divide both sides by 3," we discover that $3 \not\equiv 5 \bmod 6$. If we look at the congruence $36 \equiv 60 \bmod 8$, we see that it says that $3 \cdot 12 \equiv 3 \cdot 20 \bmod 8$; and in this case, we have $12 \equiv 20 \bmod 8$. The following theorem explains the difference between these two cases.

Proposition 3: Suppose that n is a positive integer and a, b, c are integers such that $ab \equiv ac \bmod n$. Then if a and n are relatively prime, we have $b \equiv c \bmod n$.

Proof. Since $ab \equiv ac \bmod n$, we have $n \mid ab - ac = a(b - c)$. Suppose a and n are relatively prime. Then we have $n \mid (b - c)$ by the Fundamental Theorem of Arithmetic.

Proposition 4: Suppose p is a prime. Then for all integers x such that $x \not\equiv 0 \bmod p$, there is an integer y such that $xy \equiv 1 \bmod p$.

Proof. That $x \not\equiv 0 \bmod p$ means $p \nmid x$. Since p is prime, $\gcd(p, x) = 1$. Using the Euclidean algorithm, we can write $\gcd(p, x)$ as a linear combination of p and x (Exercise 4.45); i.e., we can find integers k, y such that $kp + yx = \gcd(p, x) = 1$. Then we have $p \mid 1 - xy$; i.e., $xy \equiv 1 \bmod p$.

Example. We will compute $\gcd(124, 380)$ using the Euclidean algorithm. (See also Exercise 4.9 of the Introduction.)

$$380 = 124(3) + 8,$$
$$124 = 8(15) + 4,$$
$$8 = 4(2) + 0.$$

So $\gcd(124, 380) = 4$. Now we will write $\gcd(124, 380)$ as a linear combination of 124 and 380. Rewriting the first two equations gives

$$8 = 380 - 124(3),$$
$$4 = 124 - 8(15).$$

Substituting the first into the second, we get $4 = 124 - (380 - 124(3))(15) = (46)(124) - (15)(380)$.

To follow the proof of Sophie Germain's Theorem, we also need some results from number theory.

Lemma 4.1: *If a, b, and c are positive integers such that $\gcd(a, b) = 1$ and $a \mid bc$, then $a \mid c$.*

(See Exercise 4.46.) The following lemma is an immediate corollary.

Lemma 4.2: *If a prime number p divides a product, then p divides one of the factors.*

Proof. We will show that if $p \mid ab$ (a, b integers), then $p \mid a$ or $p \mid b$. (We leave as an exercise the argument that if

$$p \mid \prod_{i=1}^{n} a_i \quad (n \geq 1),$$

then $p \mid a_i$ for some i, $1 \leq i \leq n$.) If $p \mid a$, we are done. If $p \nmid a$, then since p is prime, we have $\gcd(p, a) = 1$. Then $p \mid b$ by Lemma 4.1.

Fundamental Theorem of Arithmetic: *Any integer $n > 1$ has a factorization*

$$n = p_1{}^{a_1} p_2{}^{a_2} \cdots p_m{}^{a_m},$$

where $p_1, \ldots, p_m$ are prime. This factorization is unique up to the order of the factors and is called the prime factorization *of n.*

Proof. To show the existence of such a prime factorization, note that if n is itself a prime, we are done. If n is not prime, then it has (positive integer) factors a, b such that $n = ab$. Again, if a and b are both prime, we are done. If a is not prime, we have factors a_1, a_2 such that $a = a_1 a_2$ (similarly for b). Note that any factor of a or b is also a factor of n. This "breaking up" process must terminate, since any positive integer n can have only a finite number of factors.

To show the uniqueness of the prime factorization, let us suppose we have two such factorizations; so that

$$n = p_1^{a_1} \cdots p_k^{a_k} = q_1^{b_1} \cdots q_l^{b_l}.$$

(We will assume that all the p_i's are distinct, and similarly for the q_j's.) Since for all i, $1 \leq i \leq k$, we have $p_i \mid q_1^{b_1} \cdots q_l^{b_l}$, by Lemma 4.2 and Exercise 4.47, we know that $p_i \mid q_j$ for some j, $1 \leq j \leq l$. Since both p_i and q_j are prime, we must have $p_i = q_j$. It follows that $k \leq l$. A similar argument shows that $l \leq k$; so $k = l$. Furthermore, the distinctness of the q_j's implies that $p_i^{a_i} \mid q_j^{b_j}$ and similarly $q_j^{b_j} \mid p_i^{a_i}$; so we must have $a_i = b_j$. Thus, the two prime factorizations are the same, up to the order of the factors.

Lemma 4.3: *If the product of two positive relatively prime numbers is an nth power, then each of the numbers is.*

Proof. Let x and y be two relatively prime numbers such that $xy = a^n$ where a is a positive integer. First we note that in the prime factorization of an nth power, all exponents must be divisible by n. For if we look at the prime factorization of $a = p_1^{a_1} \cdots p_k^{a_k}$, then $a^n = (p_1^{a_1} \cdots p_k^{a_k})^n = p_1^{na_1} \cdots p_k^{na_k}$. Clearly, if in the prime factorization of a number all exponents are divisible by n, then the number is an nth power.

The primes in the prime factorization of $xy = a^n = p_1^{na_1} \cdots p_k^{na_k}$ are exactly those in the prime factorizations of x and y. That is, each p_i must appear in the prime factorization of x or y, and each prime in the factorization of x or y is one of the p_i. Since x and y are relatively prime, if p_i is a factor of x for some i, then we must have $p_i^{na_i}$ is a factor of x (but of course, $p_i^{na_i+1} \nmid x$). So $p_i^{na_i}$ is a factor in the prime factorization of x. Similarly, if p_i were instead a factor of y, then $p_i^{na_i}$ would be a factor in the prime factorization of y. Thus, every exponent in the prime factorizations of x and y is divisible by n; so x and y are nth powers.

Exercise 4.42: Look at the definitions of congruence arithmetic in Gauss's *Disquisitiones Arithmeticae*.

Exercise 4.43: Verify Proposition 1.

Exercise 4.44: Look up the definition of an equivalence relation. Give examples of equivalence relations.

Exercise 4.45: Let d be the greatest common divisor of two positive numbers n and m. Use the Euclidean algorithm to show that there exist integers r and s such that $rn + sm = d$.

Exercise 4.46: Prove Lemma 4.1.

Exercise 4.47: Complete the proof of Lemma 4.2 by showing that if p is a prime and

$$p \mid \prod_{i=1}^{n} a_i,$$

then $p \mid a_i$ for some i, $1 \leq i \leq n$.

CHAPTER 5

Algebra: The Search for an Elusive Formula

5.1 Introduction

On the night of February 12, 1535, Niccolò Tartaglia, of Brescia, found the solution to the following vexing problem: "A man sells a sapphire for 500 ducats, making a profit of the cube root of his capital. How much is his profit?" This triumph helped him gain victory over Antonio Maria Fiore, who had posed this problem to challengers as part of a public contest. Besides fame, the prize for the winner included thirty banquets prepared by the loser for Tartaglia and his friends. (Tartaglia chose to decline this part of the prize.) Such contests were part of academic life in sixteenth-century Italy, as competitors vied for university positions and sponsorship from the nobility [93, p. 329].

In order to find a solution to the above problem Tartaglia needed to solve the equation

$$x^3 + x = 500,$$

where x is an unknown real number. The method to solve cubic equations of this type had been discovered by Scipione del Ferro, professor at the University of Bologna, who had passed on the secret to his pupil Fiore. Del Ferro's solution for cubic equations of the form $x^3 + cx = d$, in modern notation, amounts to the formula

$$x = \sqrt[3]{\sqrt{(d/2)^2 + (c/3)^3} + d/2} - \sqrt[3]{\sqrt{(d/2)^2 + (c/3)^3} - d/2}.$$

Before continuing, the reader is invited to verify that this formula, with $c = 1$, $d = 500$, does indeed produce an exact solution to the contest equation, which had given so much trouble to Tartaglia. (Note that Tartaglia did not have a calculator.) A further list of problems that Fiore posed to Tartaglia can be found in [58, p. 254].

PHOTO 5.1. Babylonian problem text on tablet YBC 4652.

Much earlier, around 2000 B.C.E., the Babylonians had solved problems that can be interpreted as quadratic equations [91, pp. 108 ff.]. In fact, many Babylonian clay tablets have been preserved with lists of mathematical problems on them. Scholars have generally assumed that the motivation for trying to solve quadratic equations originally arose from the need to solve practical or scientific problems. Alternative interpretations are suggested in [93, p. 34].

Following is a typical example of such a problem, taken from [173, p. 63] (see also [128]). The numbers, in the sexagesimal (base 60) system the Babylonians used, are given in the notation of [128]. The number $63\frac{1}{2}$, for instance, is represented as 1,3;30, where the semicolon plays the role of the decimal point. Likewise, $1; 3, 30 = 1\frac{7}{120}$. See [119, pp. 162 ff.] for a discussion of the Babylonian number system. We follow [173, pp. 63 ff.] in the interpretation of the tablet that is numbered AO 8862, dating from the Hammurabi dynasty, ca. 1700 B.C.E.

Length, width. I have multiplied length and width, thus obtaining the area. Then I added to the area, the excess of the length over the width: 3, 3 (i.e. 183 was the result). Moreover, I have added length and width: 27. Required length, width and area.

(given:) 27 and 3, 3, the sums
(result:) 15 length   3, 0 area.
                    12 width

One follows this method:

$$27 + 3, 3 = 3, 30$$

$$2 + 27 = 29.$$

Take one half of 29 (this gives 14; 30).

$$14; 30 \times 14; 30 = 3, 30; 15$$
$$3, 30; 15 - 3, 30 = 0; 15.$$

The square root of 0; 15 is 0; 30.

$$14; 30 + 0; 30 = 15 \quad \text{length}$$
$$14; 30 - 0; 30 = 14 \quad \text{width}.$$

Subtract 2, which has been added to 27, from 14, the width. 12 is the actual width. I have multiplied 15 length by 12 width.

$$15 \times 12 = 3, 0 \quad \text{area}$$
$$15 - 12 = 3$$
$$3, 0 + 3 = 3, 3.$$

Translated into modern notation, the problem amounts to solving a system of two equations

$$xy + x - y = 183,$$
$$x + y = 27.$$

The author makes a change of variable to transform the system into the standard form in which such problems were solved. If we set

$$y' = y + 2,$$

then the system gets transformed into

$$xy' = 210,$$
$$x + y' = 29.$$

The solution of a general system of the form

$$xy' = b,$$
$$x + y' = a,$$

follows a standard recipe, given by the three equations

$$w = \sqrt{(a/2)^2 - b},$$
$$x = (a/2) + w,$$
$$y' = (a/2) - w.$$

Rather than giving the general solution method in this way, it was indicated by a series of examples, all solved in this fashion.

Observe that solving our system $xy' = b, x + y' = a$ is equivalent to solving the equation $(a - x)x = ax - x^2 = b$. The Babylonians thus had a method to solve certain quadratic equations. Even some cubic and quartic equations, as well as some systems with as many as ten unknowns, were within their reach. (See also Exercises 5.1 and 5.2.) The solutions were phrased in essentially numerical terms, without recourse to geometry (in contrast to later Greek mathematics), even though quantities were represented as line segments, rectangles, etc. Also, the solutions were given in the form of a procedure, or algorithm, without any indication as to why the procedure provided the correct solution.

Beginning with the Pythagoreans (approx. 500 B.C.E.), a class of problems known as "application of areas" appeared in Greek mathematics. At around the same time, Chinese mathematics was developing elaborate methods for solving simultaneous systems of linear equations [91, pp. 173 f.][116, pp. 124, 249 ff.], and Indian Vedic mathematics developed accurate methods of calculating roots, and considered, but did not solve, certain quadratic equations [91, p. 273].

The most important type of Greek application of areas problem was the following: Given a line segment, construct on part of it, or on the line segment extended, a parallelogram equal in area to a given rectilinear figure and falling short (in the first case) or exceeding (in the second case) by a parallelogram similar to a given parallelogram [97, p. 34]. One may view the systems of equations solved by the Babylonians as application of areas problems. Indeed, suppose we want to solve the system

$$x + y = a,$$
$$xy = b^2.$$

Using the terminology of Figure 5.1, we wish to apply to $AB(= a)$ a rectangle $AH(= ax - x^2)$ equal to the given area b^2, and falling short of $AM(= ax)$ by the square BH. This is equivalent to solving the quadratic equation

$$x(a - x) = b^2.$$

The first original source we consider in this chapter is Proposition 5 from Book II of Euclid's *Elements*, written around 300 B.C.E. Together with the Pythagorean Theorem, it gives a method to solve such quadratic equations, that is, to construct the solutions with straightedge and compass. In the general statement of application of areas problems above, the "falling short" case was termed *ellipsis* and the "exceeding" case *hyperbolē*. Here lies the origin of the terminology used for the conic sections [97, pp. 91–92].

Euclid does not explicitly treat quadratic equations. Book II is a rather short part of the *Elements* and contains a collection of geometric propositions, which Zeuthen [181] calls "geometric algebra." A typical example is the following (Proposition 1): *If there be two straight lines, and one of them be cut into any number of segments whatever, the rectangle contained by the two straight lines is equal to the rectangles contained by the uncut straight line and each of the segments.* Translated into algebraic notation, this corresponds to the formula

$$a(b + c + \cdots) = ab + ac + \cdots.$$

In essence, Book II might be viewed as the beginning of a textbook on algebra, written in geometrical language, and as a continuation of Babylonian algebra [173,

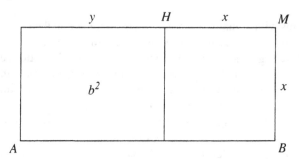

FIGURE 5.1. Application of areas.

p. 119]. Later, the Arab scholars Ibn Qurra (908–946) and Abu Kamil (ca. 850–930) used results from Book II to solve equations, illustrating the correspondence between algebra and geometry [12, Ch. 4].

Ironically, one reason for the lack of progress on higher-degree equations was the very rigor that so distinguishes classical Greek mathematics. The restriction of rigorous mathematics to geometry forced the use of geometric methods with their inherent complications and limitations. It also made it impossible to consider equations of degree higher than three in the absence of a geometric interpretation. Major progress in the theory of equations was made during the "Silver Age" of classical Greek mathematics (about 250–350 C.E.) by Diophantus of Alexandria. He systematically introduced symbolic abbreviations for the terms in an equation, while the Egyptians and Babylonians used prose. In addition, he was the first to consider exponents higher than the third. For more details see [93, Sec. 5.2].

Another reason for the lack of further progress in solving algebraic equations of degree higher than two was complications involved in the development of a rigorous treatment of irrational numbers as mathematical objects, exemplified by Books V and X of the *Elements*. It was known that not all geometric magnitudes are rational, such as the diagonal in a square of side length one (Exercise 5.3). To avoid dealing with the nature of such irrational magnitudes Euclid employs geometric algebra with elaborate proportionality arguments [97, pp. 173 ff.], [169, pp. 15–16]. Although solutions to certain cubic equations arise in the work of Archimedes (287–212 B.C.E.) and Menaechmus (middle of the fourth century B.C.E.) (see [93, pp. 108–109]), it was left to later cultures to make significant progress toward methods for solving higher-degree equations.

A first important step was accomplished by mathematics in India, during the period 200–1200 C.E. Less rigorous than Greek mathematics and much less tied to geometry, it allowed a correct arithmetic of negative and irrational numbers. This approach prevented the obsession with philosophical difficulties that prevented much progress in its Greek predecessor [97, pp. 184 ff.]. (For more details see Chapter 6 of [93].)

The rise of Islam during the seventh century and its spread from the Arabian peninsula all the way to countries bordering China in one direction and to Spain in the other direction prepared the ground for important developments in our story.

In the eighth century, the newly built city of Baghdad, the present-day capital of Iraq, became one of the political and scientific centers of the Islamic world. In a setting described in the tales of *The Thousand and One Nights*, the Caliph Harun al-Rashid had a large library constructed, containing many scientific works written in Sanskrit, Persian, and Greek. During the ninth century, the Caliph al-Ma'mun supplemented this library with a new intellectual center called the House of Wisdom. Devoted to research and a massive translation project, it led to an influx of scholars from all parts of the Islamic sphere of influence, as well as a growing collection of original manuscripts together with translations into Arabic. By the end of the century, the major works of Euclid, Archimedes, Apollonius, Diophantus, and other Greek mathematicians were available to Islamic scholars, in addition to many manuscripts of Babylonian and Indian origin. (See [12] for a detailed description of Islamic mathematics.)

One of the scholars in the House of Wisdom was al-Khwarizmi (ca. 800–847), who around 825 wrote *The Condensed Book on the Calculation of al-Jabr and al-Muqābala*. He explains the purpose of the treatise in the introduction:

> That fondness for science, by which God has distinguished the Imam al-Ma'mun, the Commander of the Faithful...has encouraged me to compose a short work on calculating by *al-jabr* and *al-muqābala*, confining it to what is easiest and most useful in arithmetic, such as men constantly require in cases of inheritance, legacies, partition, law-suits, and trade, and in all their dealings with one another, or where measuring of lands, the digging of canals, geometrical computation, and other objects of various sorts and kinds are concerned [93, p. 229].

From the title of this work derives our word *algebra* to denote the branch of mathematics that Europe later learned from it. The following interpretation of the words in the title is given in [93, p. 228].

> The term *al-jabr* can be translated as "restoring" and refers to the operation of transposing a subtracted quantity on one side of an equation to the other side where it becomes an added quantity. The word *al-muqābala* can be translated as "comparing" and refers to the reduction of a positive term by subtracting equal amounts from both sides of the equation. For example, converting $3x + 2 = 4 - 2x$ to $5x + 2 = 4$ is an example of *al-jabr* while converting the latter to $5x = 2$ is an example of *al-muqābala*.

The text contains a classification of different types of quadratic equations, with only positive coefficients. Subsequently, a collection of numerical recipes is given to solve the different types. Influenced by Greek tradition, al-Khwarizmi then gives geometric proofs for his procedures.

In studying the Greek geometry texts, one encounters problems that lead to cubic equations. The problem of "doubling the cube" was one such instance. This was the problem of doubling the cubical altar at Delos, reportedly posed by an oracle, for which the Delians sought Plato's help [93, p. 42]. The correct interpretation, according to Plato (427–348/347 B.C.E.), was to construct with straightedge and compass a cube that has twice the volume of the altar cube. If we assume the side length of the altar cube to be 1, then we are led to the problem of constructing the cube root of 2, which was shown to be impossible only in the nineteenth century by Pierre Wantzel (1814–1848) [93, p. 598]. An excellent reference for more details on these problems is [98].

If in addition to straightedge and compass one allows the use of conic sections, such as parabolas and hyperbolas, however, then problems such as this one have ready solutions. See [93, pp. 108–9]. See also [12, Ch. 3] for Islamic contributions to constructibility problems. Several Islamic mathematicians applied conic sections to other cubic equations as well. The most systematic effort in this direction is due to Omar Khayyam (1048–1131), who classified all types of cubic equations and provided solutions for them, using conic sections. In his work *Treatise on Demonstrations of Problems of al-Jabr and al-Muqabala*, he emphasizes that the reader needs to be thoroughly familiar with the work of Euclid, Apollonius, and al-Khwarizmi in order to follow the solutions. For instance, a solution to the equation $x^3 + cx = d$ is found by intersecting a circle and a parabola [93, p. 244]. An English translation of part of the *Treatise* can be found in [122].

In Persia, Khayyam's treatise was used as a school text for centuries. In the Western world Khayyam became known for his poetry through the 1859 publication of an English translation of *Rubaiyat of Omar Khayyam*. This book went through more than 300 editions and spawned a plethora of secondary works. The year 1900 saw the foundation of the Omar Khayyam Club of America in Boston, and there is continued interest in his poetry [122, pp. 583–584].

Despite Khayyam's ingenuity, his work did not lead to progress in the theory of equations, since his methods relied heavily on geometry. The world still had to wait for an algebraic method applicable to higher-degree equations. The time was ripe for such a method in sixteenth century Renaissance Italy. A number of mathematicians were working on algebraic methods for solving cubic equations. Still lacking a workable concept of negative numbers, the usual categorization was used depending on which side of the equation contained which terms (with nonnegative coefficients). As mentioned at the beginning of this section, it was Scipione del Ferro (1465–1526), a professor at the University of Bologna, who first found an algebraic method to deal with the equation $x^3 + cx = d$ (Exercises 5.4–5.5). Niccolò Tartaglia (1499–1557) claimed to possess a solution for the equation $x^3 + bx^2 = d$. This caught the attention of a Milanese mathematician, Girolamo Cardano (1501–1576). Cardano was working on an arithmetic text, and invited Tartaglia to tell him about his methods, to be included in the book, with full credit to Tartaglia. Finally, according to Tartaglia himself, Tartaglia was willing to tell Cardano what his "rule" was for finding the solutions of his cubic, but not his "method," which would explain why the rule indeed produced the right solutions. In modern terms, he was willing to give Cardano the algorithm to compute the solutions, without any indication why the algorithm indeed produced the desired result, nor how one could discover the algorithm. Cardano swore not to divulge Tartaglia's rule. However, he began working on the problem himself afterwards, and published his solution in his greatest work, the *Ars Magna* (The Great Art)

[29]. There followed a long public dispute with Tartaglia over Cardano's supposed breach of confidence, told in some detail in [93, Ch. 9]. The second text we will study in this chapter is the solution of one of the cases of the cubic out of the *Ars Magna*, the equation $x^3 = bx + d$, or, in Cardano's words, the case of "the cube equal to the first power and number."

While most of the *Ars Magna* is devoted to solving the cubic, it also contains a method for solving equations of degree 4. This method had been discovered by Ludovico Ferrari (1522–1565), a student of Cardano. Cardano included Ferrari's solution in his book but relegated it to the last of its forty chapters. As he says about it in the introduction:

> Although a long series of rules might be added and a long discourse given about them, we conclude our detailed consideration with the cubic, others being merely mentioned, even if generally, in passing. For as *positio* [the first power] refers to a line, *quadratum* [the square] to a surface, and *cubum* [the cube] to a solid body, it would be very foolish for us to go beyond this point. Nature does not permit it [29, p. 9].

Ferrari had proceeded by first making a change of variable in the general fourth-degree equation in order to eliminate the cubic term. If our equation is

$$x^4 + ax^3 + bx^2 + cx + d = 0,$$

then the substitution

$$y = x + \frac{a}{4}$$

results in the equation

$$y^4 + py^2 + qy + r = 0,$$

for the appropriate p, q, r. Now rewrite this equation as

$$\left(y^2 + \frac{p}{2}\right)^2 = -qy - r + \left(\frac{p}{2}\right)^2.$$

Adding a quantity u to the equation inside the parentheses on the left-hand side, we obtain

$$\left(y^2 + \frac{p}{2} + u\right)^2 = -qy - r + \left(\frac{p}{2}\right)^2 + 2uy^2 + pu + u^2.$$

Now, Ferrari's clever idea was to determine a constant u, depending on p and q, but not on y, that would make the right-hand side a perfect square of a first degree polynomial in y. This leads to the cubic equation

$$8u^3 + 8pu^2 + (2p^2 - 8r)u - q^2 = 0,$$

which can now be solved by Cardano's method (Exercise 5.6). This equation is now called a *resolvent cubic* for the quartic we started with, and we will have occasion to return to it in the work of Lagrange in the eighteenth century.

The formulas for the roots of third- and fourth-degree equations raised as many questions as they answered, the hallmark of any important mathematical discovery.

Just how many roots did an equation have? Why did Cardano's formula produce some solutions and not others? For instance, the equation $x^3 + 16 = 12x$ has the solution $x = 2$, but Cardano's formula produces another solution, $x = -4$. And, most importantly, why was it sometimes necessary to extract square roots of negative numbers to evaluate the formula, even when the end result was a real number? Finally, one was forced to take negative and imaginary numbers seriously as mathematical objects. Before, they had been dismissed as absurd and unnatural. It was not too long before the algebra of complex numbers appeared in the influential text *Algebra* by Rafael Bombelli (1526–1573), published in 1572.

Other areas of mathematics, such as analytic geometry and the differential calculus, began to take center stage at the beginning of the seventeenth century, and further progress in the theory of algebraic equations was slow in coming. Cardano's work pushed the rudimentary system of algebraic notation to its utmost limit, and real progress required a more user-friendly symbolism, which was slow in developing. First steps in this direction were taken by François Viète (1540–1603). (See also the number theory chapter.)

After such great success with cubic and quartic equations, the next obvious step was to attack the general quintic equation. Here progress was to be slow, however. While several new methods were developed for cubic and quartic equations, none of them resulted in a method for the quintic. The next milestone that we will look at does not appear until the second half of the eighteenth century, in the work of Joseph Louis Lagrange (1736–1813).[1] Lagrange's 1771 memoir *Réflexions sur la Résolution Algébrique des Equations* (Reflections on the Solution of Algebraic Equations) [101] is one of the gems to be found in this subject and was crucial for the eventual solution of the problem. Algebra was synonymous with the theory of equations, and Lagrange's work is the outstanding contribution to the subject during the second half of the eighteenth century. In a leisurely style, the eminently readable and lengthy work first gives an analysis of all the work done on the problem, surveys the different methods of Cardano, Euler, etc., and attempts to extract from them general principles:

> These methods all come down to the same general principle, knowing how to find functions of the roots of the given equation, which are such $1°$ that the equation or equations by which they are given, i.e., of which they are roots (equations which are commonly called the reduced equations), themselves have degree less than that of the given equation, or at least are factorable into other equations of degree less that that; $2°$ that one can easily compute from these the values of the desired roots [94, p. 45].

He makes it clear that any such method will involve in an essential way the study of certain types of permutations of the roots of the equation, a conviction borne

---

[1] Lagrange's contributions to mathematics are woven into the strands of two other chapters in this book as well. He taught an analysis course at the Ecole Polytechnique, through which he discovered that one of his pupils who submitted particularly brilliant homework assignments had them in fact done by Sophie Germain, who later made important contributions to Fermat's Last Theorem. See the number theory chapter.

out by the later course of events. (A permutation of the roots of an equation can be thought of simply as a shuffle, or rearrangement, of the roots.) The memoir contains a number of results about such permutations, which foreshadow the later emergence of group theory.[2] After many very insightful observations, he concludes with a rather pessimistic assessment of the possibility for finding a general algebraic method for solving equations of all degrees. Here we will study extensive excerpts from Lagrange's memoir to illustrate these points.

The next major advance had to wait until the emergence of new generations of mathematicians toward the beginning of the nineteenth century. First there was the Italian Paolo Ruffini (1765–1822), who in 1799 published a lengthy treatise *General Theory of Equations* [146, v. 1, pp. 1–324]. It contained a proof that the general equation of degree five was not solvable by radicals, meaning that there could be no formula for the roots that involved only algebraic operations on the coefficients of the equation together with radicals of these coefficients. Due to the less than clear exposition, his arguments were received with much skepticism (later on a significant gap was found), and were never accepted by the mathematical community [169, pp. 273 ff.].

PHOTO 5.2. Abel.

---

[2] A group is a set, together with a binary operation that combines two group elements to give a third. This operation is required to be associative, have an identity element, and each element has an inverse. An example is the set of integers with addition as the binary operation. Group theory is the study of such structures with the ultimate goal of creating a classification scheme.

In 1824, the Norwegian mathematician Niels Henrik Abel (1802–1829) gave a different proof, published in the first issue of *Journal für die Reine und Angewandte Mathematik* in 1826 [1, vol. 1, pp. 66–94]. There is some evidence [67] that Abel had similar results for higher-degree equations of prime degree, as well as results on the roots of equations that are solvable by radicals. In addition, he achieved a characterization of a large class of equations that in fact do admit a solution by radicals. This class includes all so-called cyclotomic equations, discussed below [169, p. 302]. Possibly only his untimely death at the age of 26 prevented him from being the one to give a complete solution to the problem of which algebraic equations are solvable by radicals. This honor was reserved for the Frenchman Evariste Galois (1811–1832), one of the most flamboyant characters in the history of mathematics.

At the tender age of 18, Galois communicated to the Academy of Sciences in Paris some of his results on the theory of equations, through one of its members, Augustin-Louis Cauchy (1789–1857). Shortly thereafter, Galois learned that a number of his results had actually been obtained by Abel, before him. Two years later, he submitted a rewritten version, *Mémoire sur les Conditions de Résolubilité des Equations par Radicaux* (Memoir on the Conditions for Solvability of Equations by Radicals), which, as it turned out much later, contained the punch line of our story. With Abel's work it had become clear that the search for a general method for solving algebraic equations by radicals was doomed. There remained the problem of characterizing those equations that *are* amenable to such a solution. The last source in this chapter consists of excerpts from Galois's memoir. They show his work to be a new kind of mathematics that would be characteristic of the nineteenth century, and which marked a gradual departure from the work of Lagrange, to which it nevertheless owed its inspiration. Galois's characterization of algebraic equations solvable by radicals forms one of the beginnings of abstract algebra, namely the theory of groups and that of fields. Most of a standard first-semester graduate course on abstract algebra is nowadays devoted to the legacy of Galois's groundbreaking work. The importance of his memoir was not recognized properly until 1843, when Joseph Liouville prepared Galois's manuscripts for publication and announced that Galois had indeed solved this age-old problem.

Even though much has been omitted in this summary account, it cannot end without mentioning the work of Carl Friedrich Gauss (1777–1855) on cyclotomic equations, published in 1801 in his great treatise *Disquisitiones Arithmeticae* (Arithmetical Investigations). Gauss's work was the culmination of another strand of our story, which will have to receive short shrift, but needs to be outlined in any account of the theory of equations.

As is often the case in mathematics, developments in other fields were ultimately to play a big role in the theory of equations as well. In 1702, the German mathematician and philosopher Gottfried Wilhelm Leibniz (1646–1716) published a paper in the scientific journal *Acta Eruditorum* with the curious title *New Specimen of the Analysis for the Science of the Infinite about Sums and Quadratures* [110]. In this paper, Leibniz discusses what we now call the method of partial fractions, in particular the possibility of decomposing a rational function, that is, a quotient of two polynomial functions, into a sum of fractions with denominators that are ei-

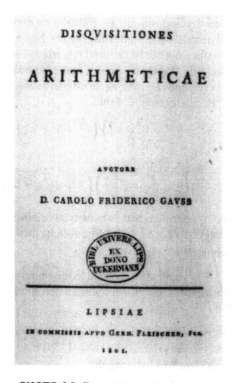

PHOTO 5.3. *Disquisitiones Arithmeticae.*

ther linear or quadratic polynomials. This would reduce the integration of rational functions to integrating only those functions involving $1/x$ and $1/(x^2 + 1)$, which Leibniz calls the "quadrature of the hyperbola and the circle"[3] (see the analysis

---

[3]In order to understand why Leibniz relates $1/(x^2 + 1)$ to the quadrature (area) of the unit circle, consider the following figure:

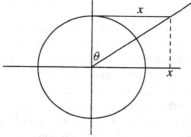

Now, using calculus, one can show that the area of the circle segment enclosed by the angle θ is given by

$$\frac{\theta}{2} = \frac{\tan^{-1}(x)}{2} = \frac{1}{2} \int_0^x \frac{dt}{t^2 + 1}.$$

chapter). To obtain such decompositions, he embarks on an investigation of the factorization of real polynomials into irreducible factors, hoping to show that every polynomial with real coefficients can be factored into irreducible factors of degree one or two. This is in fact true, and is now known as the "Fundamental Theorem of Algebra." Leibniz, however, ends up making an embarrassing mistake and comes up with the following "counterexample." From

$$x^4 + a^4 = \left(x^2 + a^2\sqrt{-1}\right)\left(x^2 - a^2\sqrt{-1}\right),$$

it follows that

$$x^4 + a^4 = \left(x + a\sqrt{-\sqrt{-1}}\right)\left(x - a\sqrt{-\sqrt{-1}}\right)\left(x + a\sqrt{\sqrt{-1}}\right)\left(x - a\sqrt{\sqrt{-1}}\right),$$

from which he erroneously concludes that no nontrivial combination of these linear factors can result in a real divisor of $x^4 + a^4$. Moreover, he thought that expressions like $\sqrt{\sqrt{-1}}$ led to a whole new type of number:

Therefore, $\int dx/(x^4 + a^4)$ cannot be reduced to the squaring of the circle or the hyperbola by our analysis above, but founds a new kind of its own [110, p. 360] [169, p. 99].

Of course, there was no need, in this case, to get into any investigations about new numbers at all. As Nikolaus Bernoulli (1687–1759) pointed out a little later, one easily solves this problem by observing that

$$x^4 + a^4 = (x^2 + a^2 + \sqrt{2}ax)(x^2 + a^2 - \sqrt{2}ax).$$

Apparently, Isaac Newton (1642–1727) had obtained the same factorization some time earlier. The mystery of extracting roots of complex numbers also was resolved eventually. In 1739, the French mathematician Abraham de Moivre (1667–1754) showed that radicals of complex numbers do not produce a new kind of number. Given an "impossible binomial $a + \sqrt{-b}$," in his terminology, he showed that if φ is an angle such that $\cos(\varphi) = a/\sqrt{a^2 + b}$, then the nth roots of $a + \sqrt{-b}$ are

$$\sqrt[2n]{a^2 + b}\left(\cos(\psi) + \sqrt{\cos^2(\psi) - 1}\right),$$

where ψ ranges over the n values $\varphi/n, (2\pi - \varphi)/n, (2\pi + \varphi)/n, (4\pi - \varphi)/n, (4\pi + \varphi)/n, \ldots$. But this expression is again of the form $a + \sqrt{-b}$ for some other values of a and b. Thus, Leibniz's objection was refuted, and subsequently serious attempts were undertaken to prove the Fundamental Theorem of Algebra. A number of proofs were given, all containing more or less serious flaws. As with several other important results, it was left to Gauss to correct all errors and fill all gaps to give the first essentially complete proof in 1799. Subsequently, he gave several different proofs of the result. An equivalent and more common formulation of it is that any nonconstant polynomial with complex coefficients factors into linear factors. (See [169, Ch. 7] for a more detailed discussion.)

Another accomplishment of Gauss needs to be mentioned, which was of great significance in validating some of Lagrange's conclusions. De Moivre's work on roots of complex numbers showed that the nth root of a number may be far from unique. A special role is assigned to the nth roots of 1, since one can easily see that if r is an nth root of a complex number s, and $\omega_1, \ldots, \omega_{n-1}$ are the distinct nth roots of 1 other than 1 itself, then the other nth roots of s are of the form $\omega_1 r, \ldots, \omega_{n-1} r$. Once again it was left to Gauss, in a brilliant piece of work published as part of the *Disquisitiones Arithmeticae*, to tell the whole story about these roots of unity, as they are commonly called. In this he built on previous work by a number of other mathematicians, notably Vandermonde, Euler, and Lagrange. The nth roots of unity are of course all roots of the polynomial $x^n - 1$, which we can factor into

$$x^n - 1 = (x - 1)(x^{n-1} + x^{n-2} + \cdots + x + 1).$$

It was shown earlier that it was sufficient to study the matter when n is prime. Gauss showed that if n is prime, then the second factor is irreducible over the rational numbers and has $n - 1$ distinct complex roots. Most importantly, these *cyclotomic equations*, as they are called, are all solvable by radicals (Exercise 5.7). Thus, Gauss had found the first instance of an important class of equations of arbitrarily high degree that are solvable by radicals. Shortly afterwards, Abel and Galois undertook the Herculean task of classifying precisely those equations that, like the cyclotomic ones, are amenable to this method of solution.

As happens often in mathematics, in the end the tools developed in order to attack the problem of solving equations by radicals took on far greater importance than the actual solution itself. We already mentioned the consequences of Galois's work for the development of abstract algebra, in particular the theory of groups and the theory of field extensions, which have become central to all of algebra as well as to applications of algebra to other subjects inside and outside of mathematics. In the case of group theory, a thorough account can be found in [179], while [48] and [174] contain excellent accounts of the historical development of Galois theory. Most importantly, Galois was advocating a paradigm shift, away from a purely computational approach, in favor of an abstract qualitative analysis of the problems of algebra. As he says in an unpublished preface to his memoir:

> Now, I think that the simplifications produced by the elegance of calculations (intellectual simplifications, I mean; there are no material simplifications) are limited; I think the moment will come where the algebraic transformations foreseen by the speculations of analysts will not find nor the time nor the place to occur any more; so that one will have to be content with having foreseen them. . . .

> Jump above calculations; group the operations, classify them according to their complexities rather than their appearances; this, I believe, is the mission of future mathematicians; this is the road on which I am embarking in this work [66, p. 9] [169, p. 397].

The next such paradigm shift was initiated by Emmy Noether (1882–1935) in the 1920s, and gave the subject of algebra the form it has today. She introduced a new level of abstraction into algebra, which led to great conceptual

PHOTO 5.4. Noether.

clarity and formal elegance, but also removed the subject somewhat from its origins. As a consequence, hardly anyone talks about equations and solution by radicals anymore when discussing Galois theory, which has become a beautiful axiomatic theory concerned with extensions of fields and the theory of groups.

Lest the reader become too complacent in marveling at the good fortune of such a complete and satisfying solution to the important and longstanding problem of solving polynomial equations, a rather rare occurrence in mathematics, there is of course still an unanswered question lingering, which opens up entirely new vistas in seemingly distant parts of mathematics. Given that the roots of an algebraic equation of degree five or greater cannot in general be expressed as algebraic functions of the coefficients of the equation, one might wonder whether there are other kinds of functions that give a representation of the roots, and if so, which ones. The fruitfulness of this question is born out by the book *Vorlesungen über das Ikosaeder* (Lectures on the Icosahedron), published in 1884 by the great nineteenth-century German mathematician Felix Klein (1849–1925) [95], in which he investigates these questions for equations of degree five. As the title suggests, unexpected connections to geometry appear, and his work has inspired much present-day research on this and related problems.

Exercise 5.1: Use the Babylonian method described in this section to solve the system

$$2xy + x - 4y = 1$$
$$x + 4y = 2.$$

What is the general principle behind the substitution to transform a given system into standard form?

Exercise 5.2: The Babylonians also solved systems of linear equations with a method similar to the one discussed in this section. Solve the following problem (see [128, p. 66]): From one field I have harvested 4 gur of grain per bùr (surface unit). From a second field I have harvested 3 gur of grain per bùr. The yield of the first field was 8,20 more than that of the second. The areas of the two fields were together 30,0 bùr. How large were the fields?

Exercise 5.3: Use the Pythagorean Theorem and the Fundamental Theorem of Arithmetic to show that the diagonal in a square of side length one cannot be a rational number.

Exercise 5.4: Use del Ferro's method to solve the following problem from Fiore's list of challenge problems that he gave to Tartaglia: "There is a tree, 12 braccia high, which was broken into two parts at such a point that the height of the part which was left standing was the cube root of the length of the part that was cut away. What was the height of the part that was left standing?"

Exercise 5.5: Use del Ferro's formula for the roots of a cubic to solve the equation $x^3 = 15x + 4$. (Note that this equation has three real roots.)

Exercise 5.6: Work out the details in Ferrari's method of solving an equation of degree four. For hints check [169, Sect. 3.2].

Exercise 5.7: Find all roots of the polynomial $x^4 - 1$.

5.2 Euclid's Application of Areas and Quadratic Equations

As mentioned in the introduction, Book II of Euclid's *Elements* is rather short, with only fourteen propositions. It contains a collection of results from "geometric algebra." And indeed a number of the results translate into well-known formulas in modern algebra. The original source we discuss in this section is Proposition 5. It turns out to be the essential ingredient in the solution of certain types of quadratic equations, aside from its use in application of areas problems, which we also discussed in the Introduction. For a more detailed discussion of Book II see [20, Sect. 7.6].

We begin with several definitions. (For a general introduction to Euclid and the *Elements* see the Euclid sections in the geometry and number theory chapters.)

PHOTO 5.5. First English edition of Euclid, 1570.

Euclid, from
Elements

BOOK II.
DEFINITIONS.

1. Any rectangular parallelogram is said to be **contained** by the two straight lines containing the right angle.
2. And in any parallelogrammic area let any one whatever of the parallelograms about its diameter with the two complements be called a **gnomon**.

The type of figure referred to as a *gnomon* in the second definition can be thought of as a square with a little corner missing (Figure 5.2). The term is also used in connection with number-theoretic problems (see [93, p. 45]). For details of the origin and use of the term "gnomon" see [51, Vol. 1, pp. 370–372].

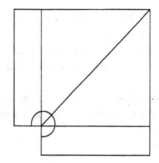

FIGURE 5.2. Gnomon.

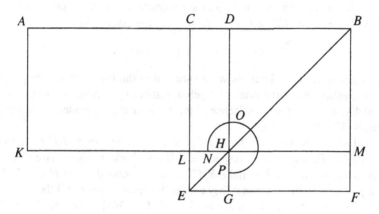

FIGURE 5.3. Proposition 5.

PROPOSITION 5.

If a straight line be cut into equal and unequal segments, the rectangle contained by the unequal segments of the whole together with the square on the straight line between the points of section is equal to the square on the half.

For let a straight line AB be cut into equal segments at C and into unequal segments at D; I say that the rectangle contained by AD, DB together with the square on CD is equal to the square on CB (Figure 5.3).

For let the square $CEFB$ be described on CB, [I. 46] and let BE be joined; through D let DG be drawn parallel to either CE or BF, through H again let KM be drawn parallel to either AB or EF, and again through A let AK be drawn parallel to either CL or BM. [I. 31]

Then, since the complement CH is equal to the complement HF, [I. 43] let DM be added to each; therefore the whole CM is equal to the whole DF.

But CM is equal to AL, since AC is also equal to CB; [I. 36] therefore AL is also equal to DF. Let CH be added to each; therefore the whole AH is equal to the gnomon NOP.

But AH is the rectangle AD, DB, for DH is equal to DB, therefore the gnomon NOP is also equal to the rectangle AD, DB.

Let LG, which is equal to the square on CD, be added to each; therefore the gnomon NOP and LG are equal to the rectangle contained by AD, DB and the square on CD.

But the gnomon NOP and LG are the whole square $CEFB$, which is described on CB; therefore the rectangle contained by AD, DB together with the square on CD is equal to the square on CB.

Therefore etc.

Q. E. D.

Translated into algebraic language, the proposition is easily understood. If we let $AC = CB = a$ and $CD = b$, then the proposition asserts that

$$(a + b)(a - b) + b^2 = a^2,$$

which is easily verified. Thus, as we mentioned in the Introduction, Proposition 5 is just another one of the rules of algebra, stated in geometric language. For a detailed discussion of the significance of this proposition the reader is encouraged to consult [173, pp. 120 ff.].

In Euclid's proof, the complement CH and the complement HF refer to the rectangles $CDHL$ and $HMFG$, respectively. If one thinks of the square $CBFE$ as containing the squares $LHGE$ and $DBMH$, then the rectangles $CDHL$ and $HMFG$ are complements in the sense that they are the figures required to fill the rest of the square $CBFE$. (See the commentary in [51, Vol. 1, p. 341] following Proposition 43 of Book I.) Note also that Euclid identifies rectangles (including squares) by opposite corners; so, for instance, DM is the square $DBMH$. The gnomon ("carpenter's square") NOP is the figure $CBFGHL$ consisting of the large square $CEFB$ with the lower left corner missing.

For our purposes, the real significance of Proposition 5 is that it gives a geometric method for solving certain quadratic equations. There is no evidence, however, that Euclid felt the same way. Suppose we wish to solve the equation

$$ax - x^2 = x(a - x) = b^2$$

for x, where $a, b > 0$. Let us further assume that $b < a/2$. We draw the picture in Figure 5.4. Then we complete it in the spirit of Euclid's Proposition 5 (Figure 5.5).

Proposition 5 applied to this picture now tells us that the area of the rectangle with sides x and $a - x$ together with the area of the small square with side length c is equal to the area of the square with side length $a/2$. In algebraic notation,

$$x(a - x) + c^2 = (a/2)^2.$$

Applying the Pythagorean Theorem to the triangle in Figure 5.5 results in

$$b^2 + c^2 = (a/2)^2.$$

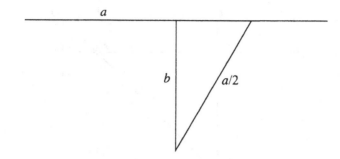

FIGURE 5.4. A quadratic equation.

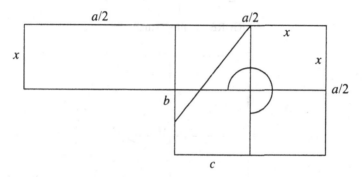

FIGURE 5.5. Solving a quadratic equation.

Combining the two equations yields

$$x(a - x) = b^2,$$

as desired.

Exercise 5.8: Find a proof of Proposition 6 in Book II in the spirit of Euclid, which says: *If a straight line be bisected and a straight line be added to it in a straight line, the rectangle contained by the whole with the added straight line and the added straight line together with the square on the half is equal to the square on the straight line made up of the half and the added straight line.* Hint: See Figure 5.6.

Exercise 5.9: Proposition 11 in Book II states: *To cut a given straight line so that the rectangle contained by the whole and one of the segments is equal to the square on the remaining segment.* Which type of quadratic equation can be solved with the help of this proposition and how?

Exercise 5.10: Read about the golden ratio (or divine proportion, as it is sometimes called) and find its connection to Proposition 11.

Exercise 5.11: Figure 5.7 accompanies the proof of Proposition 11. Here, *AB* represents the given straight line. The cutting point *H* is constructed by Euclid as

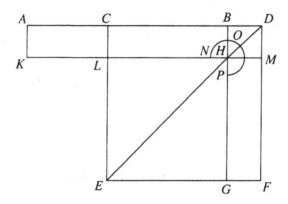

FIGURE 5.6. Proposition 6.

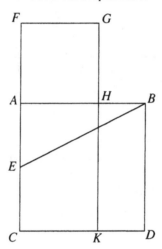

FIGURE 5.7. Proposition 11.

follows. First construct the square $ABDC$. Then bisect AC at E, and join E to B. Now extend CA until F, so as to make $EF = BE$. Construct the square $FGHA$ on AF, and extend GH to K.

Now complete the proof of Proposition 11 in the spirit of Euclid.

5.3 Cardano's Solution of the Cubic

Much of sixteenth-century Europe was in great turmoil. Wars and diseases such as the plague decimated the population and caused constant social and political upheavals. Nonetheless, it was a century of great artistic, literary, and scientific accomplishments. One of its renowned scholars was the Italian Girolamo Cardano. Known and sought after throughout Europe for his skills as a physician, he

HIERONYMI CAR
DANI, PRÆSTANTISSIMI MATHE
MATICI, PHILOSOPHI, AC MEDICI,
ARTIS MAGNÆ,
SIVE DE REGVLIS ALGEBRAICIS,
Lib.unus. Qui & totius operis de Arithmetica, quod
OPVS PERFECTVM
inscripsit,est in ordine Decimus.

H Abes in hoc libro,studiose Lector,Regulas Algebraicas (Itali, de la Cof
sa uocant) nouis adinuentionibus, ac demonstrationibus ab Authore ita
locupletatas,ut pro pauculis antea uulgó tritis iam septuaginta euaserint.Ne-
q folum , ubi unus numerus alteri,aut duo uni,uerum etiam,ubi duo duobus,
aut tres uni æquales fuerint,nodum explicant.   Hunc aût librum ideo seor-
sim edere placuit,ut hoc abstrusissimo, & plane inexhausto totius Arithmeti
cæ thesauro in lucem eruto, & quasi in theatro quodam omnibus ad spectan
dum exposito, Lectores incitarêtur,ut reliquos Operis Perfecti libros, qui per
Tomos edentur,tanto auidius amplectantur,ac minore fastidio perdiscant.

PHOTO 5.6. *Ars Magna*, original 1545 edition.

also distinguished himself as a natural philosopher, mathematician, and astrologer. Among his many books, one of the most wellknown today is his treatise on algebra, *Ars Magna* (The Great Art). However, his mathematical writings comprise only a small part of the ten volumes of his collected works, published in Lyons in 1663, under the title *Opera Omnia Hieronymi Cardani, Mediolanensis*. Cardano was a typical representative of the Renaissance. To quote from a biography:

> To Cardano, his scientific work is a means of understanding the world and human nature in general, as well as a way of gaining self-knowledge. He looks to science as an aid to orienting himself in a complex and dangerous world while being unable to effect any change. He wanted this analysis of the world and of himself to be scientific and rational; at the same time, he wanted it to satisfy the imagination as well as offer a way to comprehend the universe as a meaningful whole [60, pp. xxi ff.].

Cardano's father was a well-educated jurist and a friend of Leonardo da Vinci. He encouraged his son to study the classics and mathematics, as well as astrology. Defying the wish of his father he decided to become a physician rather than a

PHOTO 5.7. Cardano.

lawyer. After receiving a doctorate in medicine from the University of Padua in 1526, he practiced medicine, first in a small town outside of Padua, later in Milan, where he simultaneously taught mathematics. By the middle of the century his fame as a physician had spread throughout Europe, and he was summoned by nobility from as far away as Scotland for medical advice. In 1543 he obtained the chair of medicine at the University of Pavia, where he had begun his medical studies 23 years earlier. In 1562 he moved to the University of Bologna, where he remained until he was imprisoned by the Inquisition, among other things for casting the horoscope of Christ. Subsequently, he was banned from teaching, prohibited from publishing, and had to spend his final years in Rome. During his lifetime, Cardano published more than 200 works on a variety of topics, most of which went through many editions, including his autobiography *De Propria Vita Liber*, written during his final years in Rome. For biographical material on Cardano see [42, 60, 29].

To Cardano, mathematics was the language in which nature was to be described. This point of view becomes fully developed in the philosophy of Galileo two generations later, and is at the heart of the so-called scientific method, pioneered by Descartes in the following century. The *Ars Magna* was to be volume X in an encyclopedia of mathematics, which Cardano never completed and of which little remains [60, p. 73]. It is a treatise on algebraic equations, containing, besides some preparatory material, complete solutions for cubic equations as well as solutions to certain types of degree-four equations. In the introduction Cardano gives his version of the genesis of the solution for the cubic, after observing that the subject of algebra originates with al-Khwarizmi.

In our own days Scipione del Ferro of Bologna has solved the case of the cube and first power equal to a constant, a very elegant and admirable accomplishment. Since this art surpasses all human subtlety and the perspicuity of mortal talent, and is a truly celestial gift and a very clear test of the capacity of men's minds, whoever applies himself to it will believe that there is nothing that he cannot understand. In emulation of him, my friend Niccolò Tartaglia of Brescia, wanting not to be outdone, solved the same case when he got into a contest with his [Scipione's] pupil, Antonio Maria Fior, and, moved by my many entreaties, gave it to me. For I had been deceived by the words of Luca Paccioli, who denied that any more general rule could be discovered than his own.[4] Notwithstanding the many things which I had already discovered, as is well known, I had despaired and had not attempted to look any further. Then, however, having received Tartaglia's solution and seeking for the proof of it, I came to understand that there were a great many other things that could also be had. Pursuing this thought and with increased confidence, I discovered these others, partly by myself and partly through Lodovico Ferrari, formerly my pupil. Hereinafter those things which have been discovered by others have their names attached to them; those to which no name is attached are mine [29, pp. 8–9].

Del Ferro (1465–1526) was a professor at the University of Bologna, and discovered an algebraic method for solving the equation $x^3 + cx = d$, which he kept secret for reasons outlined in the introduction to the chapter until he disclosed it shortly before his death to Fior and his successor at Bologna, Annibale della Nave.

Reportedly, Tartaglia divulged his solution to Cardano in the form of a poem, after Cardano had sworn not to publish it in the book he was preparing at the time, since Tartaglia was planning on doing so himself.

When the cube and its things near
Add to a new number, discrete,
Determine two new numbers different
By that one; this feat
Will be kept as a rule
Their product always equal, the same,
To the cube of a third
Of the number of things named.
Then, generally speaking,
The remaining amount
Of the cube roots subtracted
Will be your desired count [93, pp. 329–330].

The reader is invited to verify that in this way one indeed obtains a formula for the solution to the equation $x^3 + cx = d$ (Exercise 5.12). Upon publication of the *Ars*

---

[4]Cardano is probably referring to Paccioli's characterization of $x^4 + bx^2 = ax$ and $x^4 + ax = bx^2$ as *impossible*.

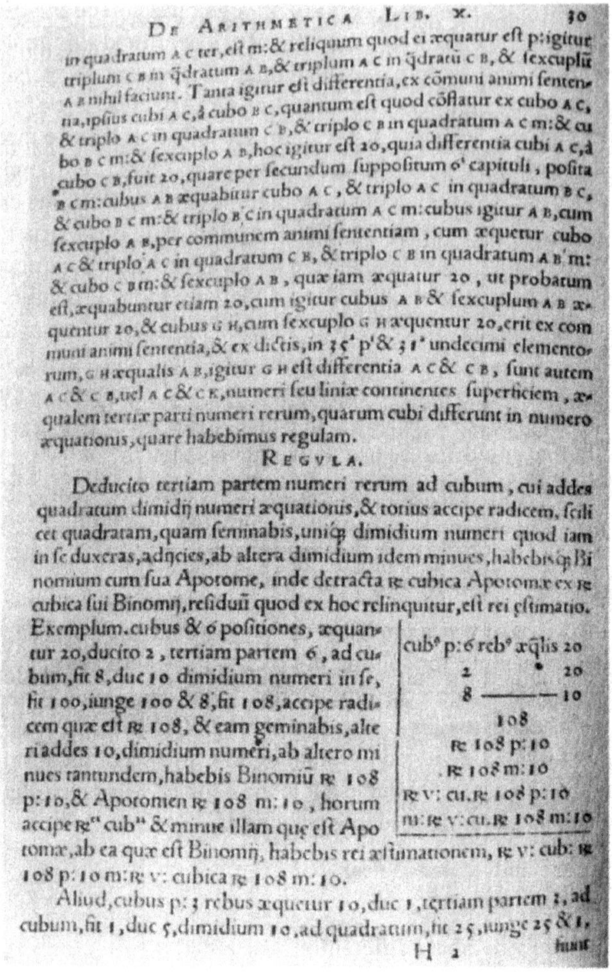

DE ARITHMETICA LIB. X.   30

in quadratum A C ter, est m:& reliquum quod ei æquatur est p:igitur
triplum c B m qdratum A B,& triplum A C in qdratu c B, & sexcuplu
A B nihil faciunt. Tanta igitur est differentia,ex cōmuni animi sententia,ipsius cubi A C,à cubo B C,quantum est quod cōflatur ex cubo A C,
& triplo A C in quadratum c B,& triplo c B in quadratum A c m:& cu
bo B c m:& sexcuplo A B,hoc igitur est 20,quia differentia cubi A c,à
cubo c B,fuit 20,quare per secundum suppositum 6ˢ capituli, posita
B c m:cubus A B æquabitur cubo A c, & triplo A c in quadratum B c,
& cubo B c m:& triplo B c in quadratum A c m:cubus igitur A B,cum
sexcuplo A B,per communem animi sententiam, cum æquetur cubo
A c & triplo A c in quadratum c B, & triplo c B in quadratum A B m:
& cubo c B m:& sexcuplo A B, quæ iam æquatur 20, ut probatum
est,æquabuntur etiam 20,cum igitur cubus A B & sexcuplum A B æ-
quentur 20,& cubus G H,cum sexcuplo G H æquentur 20,erit ex com
muni animi sententia,& ex dictis,in 3ſ' p'& 31' undecimi elemento-
rum, G H æqualis A B,igitur G H est differentia A c& c B, sunt autem
A c& c B,uel A c& c H,numeri seu liniæ continentes superficiem, æ-
qualem tertiæ parti numeri rerum,quarum cubi differunt in numero
æquationis,quare habebimus regulam.

REGVLA.

Deducito tertiam partem numeri rerum ad cubum, cui addes
quadratum dimidij numeri æquationis,& totius accipe radicem, scili
cet quadratam,quam seminabis,uniȷȵ dimidium numeri quod iam
in se duxeras,adȷȷcies,ab altera dimidium idem minues,habebisȷȷ Bi
nomium cum sua Apotome, inde detracta ℞ cubica Apotomæ ex ℞
cubica sui Binomij,residuii quod ex hoc relinquitur,est rei æstimatio.

Exemplum.cubus & 6 positiones, æquan-
tur 20,ducito 2 , tertiam partem 6 , ad cubum,fit 8,duc 10 dimidium numeri in se,
fit 100,iunge 100 & 8,fit 108,accipe radicem quæ est ℞ 108, & eam geminabis,alte
ri addes 10,dimidium numeri,ab altero mi
nues tantundem,habebis Binomiu ℞ 108
p:10,& Apotomen ℞ 108 m: 10 , horum
accipe ℞ʼcub" & minue illam quæ est Apo
tomæ,ab ea quæ est Binomij, habebis rei æstimationem, ℞ v: cub: ℞
108 p:10 m:℞ v: cubica ℞ 108 m:10.

cub⁸ p:6 reb⁸ æqlis 20

2		20
8	——	10
	108	
	℞ 108 p:10	
	℞ 108 m:10	
℞ v: cu.℞ 108 p:10		
m:℞ v:cu.℞ 108 m:10		

Aliud,cubus p:3 rebus æquetur 10,duc 1,tertiam partem 3, ad
cubum,fit 1,duc 5,dimidium 10,ad quadratum,fit 25,iunge 25 & 1,
fiunt

H 2

PHOTO 5.8. Page from the *Ars Magna* on the equation $x^3 + cx = d$.

Magna, Tartaglia publicly accused Cardano of cheating him out of the fruits of
his labor, and a lengthy argument ensued. Cardano, through his student Ludovico
Ferrari, denied ever having taken such an oath. A detailed account can be found
in the Foreword to [29]. As with most disputes of this nature it is difficult to sort
out the truth. The following observations might be useful.

In order to form any opinion on this matter one should take into account that
Tartaglia was a man of obscure origin; not even his family name is known.
"Tartaglia" means "the stutterer" and is a nickname. When the French pil-
laged Brescia in 1512, his mother sought refuge for her son in the church.
But the soldiers also invaded the sanctuary, and the twelve-year-old boy was

severely wounded by a sword cut: his jawbone was split, causing permanent damage. With enormous energy—and for the most part autodidactically—he worked to become a respected mathematics teacher and mechanical crafts-man. He lived by his skills and regarded his knowledge as his personal property. He was not a scholar in the true sense, but he had great practical knowledge—in applied mathematics as well as in mechanics—and he was a very successful teacher. Cardano, on the other hand, was a highly esteemed physician as well as a universal scholar, although he had not yet come to fame. The two men were of totally different mentalities [60, pp. 8–9].

Here we examine in detail Chapter XII of the *Ars Magna* (The Great Art), the case "On the Cube Equal to the First Power and Number," or $x^3 = cx + d$. (Note that this case differs from the case $x^3 + cx = d$, discussed above).

Girolamo Cardano, from

The Great Art

On the Cube Equal to the First Power and Number
Demonstration

Let the cube be equal to the first power and constant and let DC and DF be two cubes the product of the sides of which, AB and BC, is equal to one-third the coefficient of x, and let the sum of these cubes be equal to the constant. I say that AC is the value of x. Now since $AB \times BC$ equals one-third the coefficient of x, $3(AB \times BC)$ will equal the coefficient of x, and the product of AC and $3(AB \times BC)$ is the whole first power, AC having been assumed to be x. But $AC \times 3(AB \times BC)$ makes six bodies, three of which are $AB \times BC$ and the other three, $BC \times AB^2$. Therefore these six bodies are equal to the whole first power, and these [six bodies] plus the cubes DC and DF constitute the cube AE, according to the first proposition of Chapter VI. The cubes DC and DF are also equal to the given number. Therefore the cube AE is equal to the given first power and number, which was to be proved.

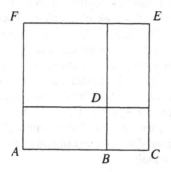

FIGURE 5.8. Cardano's demonstration.

It remains to be shown that $3AC(AB \times BC)$ is equal to the six bodies. This is clear enough if I prove that $AB(BC \times AC)$ equals the two bodies $AB \times BC^2$ and $BC \times AB^2$, for the product of AC and $(AB \times BC)$ is equal to the product of AB and the surface BE — since all sides are equal to all sides — but this [i.e., $AB \times BE$] is equal to the product of AB and $(CD + DE)$; the product $AB \times DE$ is equal to the product $CB \times AB^2$, since all sides are equal to all sides; and therefore $AC(AB \times BC)$ is equal to $AB \times BC^2$ plus $BC \times AB^2$, as was proposed.

Rule

The rule, therefore, is: When the cube of one-third the coefficient of x is not greater than the square of one-half the constant of the equation, subtract the former from the latter and add the square root of the remainder to one-half the constant of the equation and, again, subtract it from the same half, and you will have, as was said, a *binomium* and its *apotome*, the sum of the cube roots of which constitutes the value of x.

For example,

$$x^3 = 6x + 40.$$

Raise 2, one-third the coefficient of x, to the cube, which makes 8; subtract this from 400, the square of 20, one-half the constant, making 392; the square root of this added to 20 makes $20 + \sqrt{392}$, and subtracted from 20 makes $20 - \sqrt{392}$; and the sum of the cube roots of these, $\sqrt[3]{20 + \sqrt{392}} + \sqrt[3]{20 - \sqrt{392}}$, is the value of x.

Again,

$$x^3 = 6x + 6.$$

Cube one-third the coefficient of x, which is 2, making 8; subtract this from 9, the square of one-half of 6, the constant of the equation, leaving 1; the square root of this is 1; this added to and subtracted from 3, one-half the constant, makes the parts 4 and 2, the sum of the cube roots of which gives us $\sqrt[3]{4} + \sqrt[3]{2}$ for the value of x.

To the modern reader, Cardano's "demonstration" of the subsequent "rule" seems cumbersome, because of the use of geometry. There are many references to Euclid's *Elements* throughout the *Ars Magna*. Later mathematicians, in particular Viète and Descartes, developed a more efficient system of algebraic notation, completely freed from geometry (see, for instance, [22, pp. 84 ff.]).

What Cardano is proving in his demonstration of the rule for the equation $x^3 = cx + d$ is that it is sufficient to find quantities u and v such that $u^3 + v^3 = d$ and $3uv = c$. Then the solution is $x = u + v$. His proof becomes quite transparent if one uses a cube with side of length x, subdivided into eight "bodies," as indicated in Figure 5.9. From such a subdivision one can immediately deduce the Binomial Theorem for exponent 3, namely

$$(u + v)^3 = u^3 + 3uv^2 + 3u^2v + v^3.$$

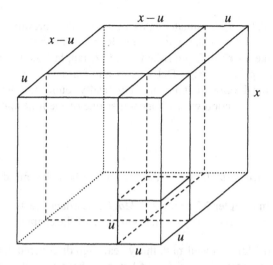

FIGURE 5.9. Cardano's cube.

This is none other than the "first proposition in Chapter VI" which he refers to. The main part of his argument is intended to show that

$$3uv^2 + 3u^2v = cx,$$

which is straightforward to reconstruct from the subdivided cube. Once he can show this, the desired formula follows easily, since he now is reduced to solving the system of equations

$$u^3 + v^3 = d,$$
$$3uv = c.$$

Its solution readily leads to the desired formula for x:

$$x = \sqrt[3]{d/2 + \sqrt{(d/2)^2 - (c/3)^3}} + \sqrt[3]{d/2 - \sqrt{(d/2)^2 - (c/3)^3}}.$$

(Why does this formula seem different from the one at the beginning of the Introduction?) Of course, if the expression under the square root in the formula is negative, that is, if $(d/2)^2 - (c/3)^3 < 0$, then one faces the awkward problem of not being able to evaluate the formula. This might not be so bad if it were to happen only with equations that have no real roots. But evaluation of Cardano's formula for the example

$$x^3 = 15x + 4,$$

which has the solution $x = 4$, leads to the expression

$$x = \sqrt[3]{2 + \sqrt{-121}} + \sqrt[3]{2 - \sqrt{-121}}.$$

Basically, Cardano was at a loss how to deal with this case and dismisses it as absurd and useless in Chapter 37, where he deals with negative square roots [29,

p. 220], [169, pp. 29–30]. Fortunately, this attitude did not prevail for long. Complex numbers were here to stay and already Rafaele Bombelli (1526–1572), in his influential algebra treatise published in 1572, taught how to calculate with cube roots of complex numbers. Del Ferro's and Cardano's formula thus gave great impetus to the development of a mathematically sound treatment of complex numbers. In the next section we will encounter one of the milestones along that road.

Exercise 5.12: Turn Tartaglia's poem into the formula in the Introduction.

Exercise 5.13: In Chapter XI of the *Ars Magna*, Cardano treats the case of the equation $x^3 + cx = d$. As he says at the beginning of the chapter:

Scipio Ferro of Bologna well-nigh thirty years ago discovered this rule and handed it on to Antonio Maria Fior of Venice, whose contest with Niccolò Tartaglia of Brescia gave Niccolò occasion to discover it. He [Tartaglia] gave it to me in response to my entreaties, though withholding the demonstration. Armed with this assistance, I sought out its demonstration in [various] forms. This was very difficult [29, p. 96].

The rule Cardano refers to reads as follows:

Cube one-third the coefficient of x; add to it the square of one-half the constant of the equation; and take the square root of the whole. You will duplicate this, and to one of the two you add one-half the number you have already squared and from the other you subtract one-half the same. You will then have a *binomium* and its *apotome*. Then, subtracting the cube root of the *apotome* from the cube root of the *binomium*, the remainder or that which is left is the value of x [29, pp. 98–99].

Find a demonstration for this rule in the spirit of Cardano. Hint: Use the picture in Figure 5.10.

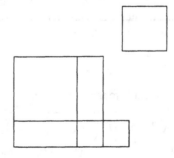

FIGURE 5.10. Cardano's rule.

5.4   Lagrange's Theory of Equations

The second half of the eighteenth century was not very favorably disposed toward pure mathematics. Since the time of Newton and Leibniz, the geometers, as mathematicians called themselves, were busily working on the development of the calculus. Euler's genius and phenomenal output defined the central problems and lines of development. The astonishing practical applications of the new theory left little time to catch one's breath and worry about the somewhat shaky foundations on which people juggled derivatives, integrals, and infinite series. There was much political and economic pressure to solve problems such as accurate navigation at sea, or construction of efficient turbines. Thus, there was neither livelihood nor prestige to be found in working on problems such as the theory of algebraic equations.

To be a professional mathematician in the eighteenth century meant to have a wealthy sponsor and be part of a scientific academy, or be independently wealthy. There was no instruction in higher mathematics at universities, leaving only private tutors if one wanted to be led to the edge of mathematical research. That is how Euler earned a living for a while, and so did several of the Bernoullis. The two leading academies during the second half of the eighteenth century were the Academies of Science in Berlin and Paris. The two men who dominated these institutions and set the mathematical agenda for all of Europe were Euler, in Berlin, and Jean d'Alembert (1717–1783), in Paris. But both men were nearing the end of their lives. (Both died in 1783, Euler at the age of 76 and d'Alembert at 66.) Who was going to take their place? (Gauss, who was destined to become the mathematical titan of the nineteenth century, was only six years old in 1783.)

Just in time, a young man from Turin, Giuseppe Lodovico Lagrangia, later changed to Joseph Louis Lagrange, caught the attention of Euler and d'Alembert. At the early age of 18 he impressed both with his communications on what was later to become the calculus of variations, and applications of it to mechanics. D'Alembert quickly adopted Lagrange as his protegé, and eventually succeeded in 1766 in making him Euler's successor at the Academy of Sciences in Berlin, after Euler left to take a position at the Academy of Sciences in St. Petersburg. Thus, Lagrange became one of the most influential mathematical scientists in Europe, especially after the deaths of d'Alembert and Euler. These men shared a serious concern for the future of mathematics, in light of strong competition from other sciences. This concern is expressed strongly in the correspondence between d'Alembert and Lagrange, with both men being rather pessimistic. As Lagrange says in a letter to d'Alembert in 1781:

> I begin to notice how my inner resistance increases little by little, and I cannot say whether I will still be doing geometry ten years from now. It also seems to me that the mine has maybe already become too deep and unless one finds new veins it might have to be abandoned.

> Physics and chemistry now offer a much more glowing richness and much easier exploitation. Also, the general taste has turned entirely in this direc-

PHOTO 5.9. Lagrange.

tion, and it is not impossible that the place of Geometry in the Academies will someday become what the role of the Chairs of Arabic at the universities is now [101, vol. 13, p. 368].

(For a more detailed discussion see [152].)

Lagrange's mathematical career can be divided into three periods. The first one was spent in his native Turin. Then, in 1766, he moved to the Berlin Academy, and finally, in 1787, he took a position at the Academy of Sciences in Paris, where he remained until his death. (For a biography of Lagrange see [42].)

Lagrange's work on the solution of algebraic equations was done while he was director of the mathematics section of the Academy of Sciences in Berlin, in the service of the Prussian king, Frederick II. Perhaps inspired by Euler's attention to algebraic problems, this work comes at a time when the most popular topic in European mathematics was infinitesimal analysis and its applications. His choice of topic is particularly remarkable, since it deals with a subject that defied any attempts at significant progress for the preceding 200 years, since the time when Cardano published the *Ars Magna*, our previous source. A number of illustrious mathematicians had tried their luck with it, resulting in a variety of methods for the algebraic solution of equations of degree three and four, but without any progress for higher degrees. No general method was in sight, not even for a significant special class of equations, until Gauss's work on cyclotomic equations at the beginning of the nineteenth century.

Lagrange's work on the theory of equations falls into two categories, one of which is the algebraic solution of polynomial equations, which for him was essentially synonymous with the subject of algebra itself, and techniques for the numerical solution of particular equations, which he views as a part of arithmetic:

> It is necessary to distinguish the solution of numerical equations from that which one calls in Algebra the general solution of equations. The first is, properly speaking, an arithmetical operation, certainly founded on the general principles of the theory of equations, but of which the results are only numbers, where one no longer recognizes the original numbers which served as elements [coefficients], and which do not conserve a trace of the different particular operations which produced them. The extraction of square and cube roots is the simplest operation of this genre; it is the solution of numerical equations of the second and third degrees with all intermediate terms missing. Also, it is convenient to give in Arithmetic the rules for solution of numerical equations, reserving for Algebra the demonstration of those which depend on the general theory of equations [101, vol. VIII, pp. 13–14], [82, p. 18].

As mentioned in the Introduction to this chapter, Lagrange's main algebraic work contained a careful analysis of known methods for solving equations of degree three and four, with the aim of finding common general principles that could be used to find methods that worked for higher equations. This memoir profoundly influenced the work of Abel and Galois, which concludes the quest for an algebraic solution to general equations stretching over two millenia. The source in this section is an excerpt from Lagrange's memoir *Réflexions sur la Résolution Algébrique des Equations* (Reflections on the Algebraic Solution of Equations), published in the Nouveaux Mémoires de l'Académie Royale des Sciences et Belles-Lettres de Berlin in 1770/71. They have been republished in [101, vol III, pp. 204–421]. Here we use pp. 206–209, 210, 213–220, 305–306, 355–356. We focus mostly on his analysis of Cardano's method for solving the cubic. This analysis is followed by some investigations of possible approaches to the general problem. In conclusion, Lagrange is rather pessimistic about the possibility of arriving at a general solution in this way.

Joseph Louis Lagrange, from
Reflections on the Algebraic Solution of Equations

In this Memoir I propose to examine the various methods for the algebraic solution of equations which have been found up to now, to reduce them to general principles, and to show *a priori* why these methods succeed for the third and fourth degree and fail for higher degrees.

This examination has a double advantage: on the one hand it serves to shed more light on the known solutions for the third and fourth degree; on the other it will be useful to those who want to occupy themselves with the solution of higher degrees, by providing them with different aspects of this subject and especially by saving them a large number of steps and futile attempts.

———

FIRST SECTION.
ON THE SOLUTION OF EQUATIONS OF DEGREE THREE.

1. Since the solution of equations of degree two is very easy, and is remarkable only by virtue of its extreme simplicity, I will begin the subject with equations of degree three which require for their solution special tricks that do not present themselves naturally.

Let

$$x^3 + mx^2 + nx + p = 0$$

therefore be the general equation of degree three. As is known, one can always make the second term of any such equation disappear by adding to its root the coefficient of the second term divided by the exponent of the first. We may therefore suppose, for the sake of greater simplicity, that $m = 0$, which will reduce the proposed equation to the form

$$x^3 + nx + p = 0.$$

It is in this form that equations of degree three have first been treated by Scipio Ferreo and by Tartaglia, to whom we owe their solution; but we shall ignore the path that they have travelled. The most natural method to arrive there seems to me to be the one that Hudde[5] devised, and which consists of representing the root as the sum of two unknowns which allows the partition of the equation into two proper parts in such a way that the two indeterminates depend only on an equation which is solvable in the same fashion as those of degree two.

Thus, according to this method one will set $x = y + z$ which, after substitution into the proposed equation will reduce it to

$$y^3 + 3y^2z + 3yz^2 + z^3 + n(y + z) + p = 0,$$

which one can put into the simpler form

$$y^3 + z^3 + p + (y + z)(3yz + n) = 0.$$

If from this equation one now produces the following two distinct equations

$$y^3 + z^3 + p = 0,$$
$$3yz + n = 0,$$

one will obtain

$$z = -\frac{n}{3y},$$

———

[5] Jan Hudde (1628–1704), Dutch mathematician.

and, by substituting into the first,

$$y^3 - \frac{n^3}{27y^3} + p = 0,$$

that is,

$$y^6 + py^3 - \frac{n^3}{27} = 0.$$

This equation is in fact of degree six, but since it contains only two different powers of the unknown, and one is the square of the other, it is clear that it can be solved like a second degree equation. In fact, one has first of all

$$y^3 = -\frac{p}{2} \pm \sqrt{\frac{p^2}{4} + \frac{n^3}{27}},$$

and then

$$y = \sqrt[3]{-\frac{p}{2} \pm \sqrt{\frac{p^2}{4} + \frac{n^3}{27}}}.$$

In this way one will know y and z and therefore will have

$$x = y + z = y - \frac{n}{3y}.$$

2. Several remarks about this solution are in order. First of all, it is clear that the quantity y can have six values, since it depends on an equation of degree six. Consequently, the quantity x also has six values. But since x is the root of an equation of degree three, one knows that it cannot have more than three distinct values. Hence it follows that the six values in question reduce to three, since each appears twice.

· · ·

3. Then, since among the six values of y, there are only three which give different values for x, we now need to distinguish these values. For this it is necessary to find the particular expression for each of the six values for y. And if one lets 1, α and β denote the three cube roots of unity, that is, the three roots of the equations $x^3 - 1 = 0$, then it is easy to see that the six values of y are as follows (abbreviating $p^2/4 + n^3/27 = q$),

$$\sqrt[3]{-\frac{p}{2} \pm \sqrt{q}}, \quad \alpha\sqrt[3]{-\frac{p}{2} \pm \sqrt{q}}, \quad \beta\sqrt[3]{-\frac{p}{2} \pm \sqrt{q}}.$$

· · ·

5. The equation of degree six

$$y^6 + py^3 - \frac{n^3}{27} = 0$$

is called the *reduced equation* for degree three, since the solution of the proposed equation

$$x^3 + nx + p = 0$$

reduces to solving this one. Now, we have already seen above how the roots of this latter equation depend on the roots of the former. Let us see conversely, how the roots of the *reduced equation* depend on the roots of the proposed equation. But to make this investigation more general and more illuminating, it is good to consider an equation which has all its terms, such as

$$x^3 + mx^2 + nx + p = 0.$$

Let the roots be represented generally by a, b, c. One proceeds now to make the second term vanish by supposing $x = x' - m/3$, and by abbreviating

$$n' = n - \frac{m^2}{3}, \qquad p' = p - \frac{mn}{3} + \frac{2m^3}{27}.$$

Then one obtains the transformed equation

$$x'^3 + n'x' + p' = 0,$$

which is of the required form. If one now sets $x' = y - n'/(3y)$, then one obtains the reduced equation

$$y^6 + p'y^3 - \frac{n'^3}{27} = 0.$$

By calling r the cube root of

$$-\frac{p'}{2} + \sqrt{\frac{p'^2}{4} + \frac{n'^3}{27}},$$

one obtains from this the three values of y, namely

$$y = r, \quad y = \alpha r, \quad y = \beta r.$$

These give the three roots

$$x' = r - \frac{n'}{3r}, \quad x' = \alpha r - \frac{n'}{3\alpha r}, \quad x' = \beta r - \frac{n'}{3\beta r}.$$

From these one has, because of $x = x' - m/3$, the three values of x, namely

$$-\frac{m}{3} + r - s, \quad -\frac{m}{3} + \alpha r - \frac{s}{\alpha}, \quad -\frac{m}{3} + \beta r - \frac{s}{\beta},$$

(abbreviating $n'/(3r) = s$), hence

$$a = -\frac{m}{3} + r - s,$$

$$b = -\frac{m}{3} + \alpha r - \frac{s}{\alpha},$$

$$c = -\frac{m}{3} + \beta r - \frac{s}{\beta}.$$

Subtracting successively the second and third of these equations from the first, one has

$$a - b = (1 - \alpha)\left(r + \frac{s}{\alpha}\right),$$

$$a - c = (1 - \beta)\left(r + \frac{s}{\beta}\right),$$

from which one obtains

$$\frac{\alpha(a - b)}{1 - \alpha} = \alpha r + s,$$

$$\frac{\beta(a - c)}{1 - \beta} = \beta r + s.$$

Subtracting once again one from the other, and subsequently dividing by $\alpha - \beta$, there results

$$r = \frac{\frac{\alpha(a-b)}{1-\alpha} - \frac{\beta(a-c)}{1-\beta}}{\alpha - \beta},$$

that is,

$$r = \frac{a}{(1 - \alpha)(1 - \beta)} + \frac{\alpha b}{(\alpha - 1)(\alpha - \beta)} + \frac{\beta c}{(\beta - 1)(\beta - \alpha)}.$$

Now, since 1, α, and β are (by hypothesis) the three roots of the equation $x^3 - 1 = 0$, one has

$$x^3 - 1 = (x - 1)(x - \alpha)(x - \beta),$$

and, differentiating,

$$3x^2 = (x - \alpha)(x - \beta) + (x - 1)(x - \beta) + (x - 1)(x - \alpha).$$

By successively setting $x = 1, \alpha, \beta$, one obtains from this

$$3 = (1 - \alpha)(1 - \beta),$$
$$3\alpha^2 = (\alpha - 1)(\alpha - \beta),$$
$$3\beta^2 = (\beta - 1)(\beta - \alpha).$$

Thus, substituting these values in the preceding expression for r, one has

$$r = \frac{a}{3} + \frac{b}{3\alpha} + \frac{c}{3\beta},$$

or, because of $\alpha\beta = 1$,

$$r = \frac{a + \beta b + \alpha c}{3}.$$

This is therefore the value of r, and consequently also of y. And since it is allowed to interchange α and β,

$$y = \frac{a + \alpha b + \beta c}{3}.$$

6. One can now see from this expression of y why the *reduced equation* necessarily has degree six. Namely, this reduced equation does not directly depend on the roots a, b, c of the proposed equation, but only on the coefficients m, n, p. Since the three roots are not distinguished from each other, it is clear that in the expression for y one is free to interchange the quantities a, b, c. Consequently, the quantity y will have as many different values as the three roots a, b, c admit permutations. Now, one knows from the theory of combinations that the number of permutations, that is, the number of different arrangements of three things, is $3 \cdot 2 \cdot 1$, hence the reduced equation in y must also have degree $3 \cdot 2 \cdot 1$, that is, degree six.

Moreover, the same expression of y shows also why the reduced equation is solvable in the same fashion as an equation of degree two, because it is clear that this follows from the fact that this equation contains only the terms y^3 and y^6, that is, the powers whose exponents are multiples of 3. Hence, if r is one of the values of y, then αr and βr also have to be, since $\alpha^3 = 1$ and $\beta^3 = 1$; but this is what happens in the expression for y which we found above. To make this easy to see we remark that $\beta = \alpha^2$. This is because, since one has $\alpha\beta = 1$ and $\alpha^3 - 1 = 0$, one also has $\alpha\beta = \alpha^3$, and therefore $\beta = \alpha^2$. Since the expression for y can be put into the form

$$y = \frac{a + \alpha b + \alpha^2 c}{3},$$

one obtains from all the possible permutations of the quantities a, b, c the following six values

$$\frac{a + \alpha b + \alpha^2 c}{3},$$
$$\frac{a + \alpha c + \alpha^2 b}{3},$$
$$\frac{b + \alpha a + \alpha^2 c}{3},$$
$$\frac{b + \alpha c + \alpha^2 a}{3},$$
$$\frac{c + \alpha b + \alpha^2 a}{3},$$
$$\frac{c + \alpha a + \alpha^2 b}{3},$$

which are therefore the six roots of the reduced equation. Now, if one multiplies the first by α, and afterwards by β or α^2, then one will have, because of $\alpha^3 = 1$, the two expressions

$$\frac{c + \alpha a + \alpha^2 b}{3} \quad \text{and} \quad \frac{b + \alpha c + \alpha^2 a}{3},$$

which are sixth and the fourth, respectively. And if one likewise multiplies the second by α and by α^2, one has

$$\frac{b + \alpha a + \alpha^2 c}{3} \quad \text{and} \quad \frac{c + \alpha b + \alpha^2 a}{3},$$

which are the third and the fifth, respectively. One has the same, if one multiplies the third and the fourth, or the fifth and the sixth by α and by α^2, because one will have the same as with all the others.

The expression of y that Lagrange refers to at the beginning of Section 6 is the reduced equation

$$y^6 + p'y^3 - \frac{n'^3}{27},$$

from which his claims follow immediately.

7. This leads us to a direct method for finding the reduced equation on which the solution of the third degree equation depends. Suppose that

$$x^3 + mx^2 + nx + p = 0$$

is the proposed equation whose roots are a, b, c, and let us suppose that the roots of the reduced equation are represented generally by a function of degree one in the roots a, b, c, such as

$$Aa + Bb + Cc,$$

where $\mathbf{A}, \mathbf{B}, \mathbf{C}$ are coefficients which are independent of the quantities a, b, c. If one carries out all possible permutations of the quantities a, b, c, one obtains the quantities

$$Aa + Bb + Cc,$$
$$Aa + Bc + Cb,$$
$$Ab + Ba + Cc,$$
$$Ab + Bc + Ca,$$
$$Ac + Bb + Ca,$$
$$Ac + Ba + Cb,$$

which are the six roots of the reduced equation. Now, in order for this equation to have only powers whose exponent is a multiple of 3, it is necessary, as we have seen above, that if r is one of the roots, then αr and βr, or $\alpha^2 r$, are also roots. Taking the quantity

$$Aa + Bb + Cc$$

for r, it follows that the quantity

$$\alpha Aa + \alpha Bb + \alpha Cc$$

is equal to one of the five other quantities below it. Now, it could not be equal to either

$$Aa + Bc + Cb$$

or

$$Ab + Ba + Cc,$$

unless $\alpha = 1$, since in the first case one would have

$$\alpha A = A,$$

and in the second

$$\alpha C = C.$$

But by comparing it to the quantity

$$Ab + Bc + Ca,$$

one has

$$\alpha A = C, \quad \alpha B = A, \quad \text{and } \alpha C = B,$$

from which one concludes that

$$C = \alpha A, \quad B = \alpha^2 A, \quad \text{and } \alpha^3 A = A,$$

that is,

$$\alpha^3 = 1.$$

This shows that α is indeed one of the roots of the equation

$$x^3 - 1 = 0.$$

If one now sets $A = 1$, for simplicity, then one has

$$A = 1, \quad B = \alpha, \quad \text{and} \quad C = \alpha^2,$$

which gives the same formulas which one found above, neglecting the denominator 3. Now, abbreviating

$$r = a + \alpha b + \alpha^2 c,$$
$$s = a + \alpha c + \alpha^2 b,$$

one obtains $r, \alpha r, \alpha^2 r$ and $s, \alpha s, \alpha^2 s$ as the six roots of the transformed equation. Now, if we call the unknown of this equation y, one will find that the product of the three factors $y - r, y - \alpha r, y - \alpha^2 r$ is equal to $y^3 - r^3$, and, likewise, the product of the other three is $y^3 - s^3$. Therefore, the product of all of them, that is, the reduced equation itself, will be represented by

$$y^6 - (r^3 + s^3)y^3 + r^3 s^3 = 0,$$

which is the required formula. The only task left to do is to find the values of $r^3 + s^3$ and $r^3 s^3$. Now, taking the cube of the quantity r, and keeping in mind that $\alpha^3 = 1$, one finds that

$$r^3 = a^3 + b^3 + c^3 + 6abc + 3\alpha(a^2 b + b^2 c + c^2 a) + 3\alpha^2(ab^2 + bc^2 + ca^2),$$

and consequently, by interchanging b and c,

$$s^3 = a^3 + b^3 + c^3 + 6abc + 3\alpha(a^2 c + c^2 b + b^2 a) + 3\alpha^2(c^2 a + b^2 c + a^2 b).$$

For simplicity, let

$$a^3 + b^3 + c^3 + 6abc = \mathbf{L},$$
$$a^2 b + b^2 c + c^2 a = \mathbf{M},$$
$$a^2 c + b^2 a + c^2 b = \mathbf{N}.$$

Then one has

$$r^3 = \mathbf{L} + 3\alpha\mathbf{M} + 3\alpha^2\mathbf{N},$$
$$s^3 = \mathbf{L} + 3\alpha\mathbf{N} + 3\alpha^2\mathbf{M},$$

hence

$$r^3 + s^3 = 2\mathbf{L} + 3(\alpha + \alpha^2)(\mathbf{M} + \mathbf{N}).$$

But since 1, α and α^2 are the three roots of the equation $x^3 - 1 = 0$, which is missing a second term, one will have that

$$1 + \alpha + \alpha^2 = 0.$$

Consequently,

$$r^3 + s^3 = 2\mathbf{L} - 3(\mathbf{M} + \mathbf{N}).$$

Now multiply together the values for r^3 and s^3 to obtain

$$r^3 s^3 = \mathbf{L}^2 + 9(\mathbf{M}^2 + \mathbf{N}^2) + 3(\alpha + \alpha^2)[\mathbf{L}(\mathbf{M} + \mathbf{N}) + 3\mathbf{MN}],$$

and hence, because of $\alpha + \alpha^2 = -1$,

$$r^3 s^3 = \mathbf{L}[\mathbf{L} - 3(\mathbf{M} + \mathbf{N})] + 9[(\mathbf{M} + \mathbf{N})^2 - 3\mathbf{MN}].$$

Now it is easy to see that the quantities $\mathbf{L}$, $\mathbf{M} + \mathbf{N}$, and $\mathbf{MN}$ are given by the coefficients m, n, p of the proposed equation, and without extraction of roots. It then follows that these quantities don't change, no matter what permutations of the quantities a, b, c one applies, since they have a unique value.

8. Indeed, given that

$$-m = a + b + c, \quad n = ab + ac + bc, \quad -p = abc,$$

one therefore has, by the known rules, that

$$a^2 + b^2 + c^2 = m^2 - 2n, \quad a^3 + b^3 + c^3 = -m^3 + 3mn - 3p.$$

From this one finds that

$$a^3b^3 + a^3c^3 + b^3c^3 = n^3 - 3mnp + 3p^2.$$

Therefore,

$$L = -m^3 + 3mn - 9p,$$
$$M + N = 3p - mn,$$
$$MN = n^3 + p(m^3 - 6mn) + 9p^2,$$

from which one finds that

$$r^3 + s^3 = -2m^3 + 9mn - 27p,$$
$$r^3s^3 = m^6 - 9m^4n + 27m^2n^2 - 27n^3 = (m^2 - 3n)^3.$$

Thus, our reduced equation is

$$y^6 + (2m^3 - 9mn + 27p)y^3 + (m^2 - 3n)^3 = 0,$$

which comes to the same thing as what we found above (5), observing however, that the unknown y here is three times the unknown y there. Let us therefore resolve this equation in the manner of those of second degree, abbreviating $y^3 = z$ for simplicity, so that one obtains

$$z^2 + (2m^3 - 9mn + 27p)z + (m^2 - 3n)^3 = 0.$$

If we call the roots of this second degree equation z' and z'', then one has

$$y^3 = z', \quad y^3 = z,''$$

hence

$$y = \sqrt[3]{z'} \text{ or } y = \sqrt[3]{z''}.$$

Consequently, since the assumption was that r and s are two values of y, one has

$$r = a + \alpha b + \alpha^2 c = \sqrt[3]{z'},$$
$$s = a + \alpha c + \alpha^2 b = \sqrt[3]{z''}.$$

These equations, combined with the equation

$$a + b + c = -m,$$

serve to find the three roots a, b, c. Indeed, because of $\alpha^3 = 1$ and $1 + \alpha + \alpha^2 = 0$, one has

$$a = \frac{-m + \sqrt[3]{z'} + \sqrt[3]{z''}}{3},$$

$$b = \frac{-m + \alpha^2\sqrt[3]{z'} + \alpha\sqrt[3]{z''}}{3},$$

$$c = \frac{-m + \alpha\sqrt[3]{z'} + \alpha^2\sqrt[3]{z''}}{3},$$

which agrees with what was proven above.

At least for exponent 3, Lagrange has now reached an important preliminary goal, namely to find a general method for producing the reduced equation in terms of the roots of the proposed equation and roots of unity. In turn, the roots of the proposed equation can be obtained from the roots of the reduced equation, which is actually solvable, even though it is an equation of degree six. Furthermore, he has now given an explanation why the reduced equation has to be of degree six, and why it ends up being a quadratic equation in y^3. That is, he has explained *a priori*, as he says at the beginning of the memoir, why the method is what it is.

The memoir now proceeds to analyze other methods for degree three equations, and he reaches the same conclusion. His next task is to analyze equations of degree four, and again he arrives at such a general method. His hope for higher-degree equations is, of course, that he can do something similar. As the next excerpt shows, this is somewhat unlikely, however.

. . .

SECTION THREE
ON THE SOLUTION OF EQUATIONS OF DEGREE FIVE AND HIGHER

The problem of the solution of equations of degree higher than four is one of those that one has not yet been able to overcome, although nobody has proven its impossibility either. Up to now I only know of two methods which could have any hope for success. Those are the one by Mr. Tschirnaus,[6] published in the *Actes de Leibsic* in 1683, and the other being the one proposed almost simultaneously by MMrs. Euler and Bezout,[7] the former in the *Nouveaux Commentaires de Petersbourg*, vol. IX, and the latter in the *Mémoires de l'Académie des Sciences de Paris* for the year 1765. These methods have the advantage of giving a solution of equations of degrees three and four in a general and uniform manner, as seen in the preceding sections, which is particular to them, and which may consequently be considered promising for their success in higher degrees. But the calculations which they require for equations of degree five and higher are so long and complicated that even the most intrepid calculator could become discouraged. Indeed, in order to apply, for example, the method of Mr. Tschirnaus to degree five, it will be necessary to solve four equations in four unknowns, of which the first one is of degree one, the second one of degree two, and so on. Thus, the final equation resulting from elimination of three of these unknowns will in general rise to degree $1 \cdot 2 \cdot 3 \cdot 4$, that is, to degree 24. Now, independent of the immense work which will be necessary to obtain this equation, it is clear that once one has found it, one will have scarcely made much progress, unless one can reduce it to a degree less than five, a reduction which, if it is possible, can only be the fruit of labors even more considerable than the first.

Following the method of M. Euler, one also necessarily arrives at a reduced equation of degree twenty-four, even though that method appears to promise a

---

[6] Ehrenfried Tschirnhaus (1651–1708), German mathematician.
[7] Etienne Bezout (1739–1783), French mathematician.

reduced equation of only degree four, by reasoning that for degree three it gives a reduced equation of only degree two, and for degree four a reduced equation of degree three. In this regard M. Bezout remarks with good reason that this is an accidental simplification which in degree four diminishes the reduced equation of M. Euler to degree three, which in general is of degree $2 \cdot 3$, that is, of degree six. And furthermore this simplification is due to the fact that the exponent 4 is a composite number.

. . .

SECTION FOUR
CONCLUSION OF THE PRECEDING REFLECTIONS WITH SEVERAL GENERAL REMARKS ON THE TRANSFORMATION OF EQUATIONS AND THEIR REDUCTION OR LOWERING TO A LESSER DEGREE

86. One has seen in the analysis we have made of the principal known methods for the solution of equations that these methods all reduce to the same general principle, namely knowing how to find functions of the roots of the proposed equation, which are such that 1° the equation or equations by which they will be given, that is, of which they are the roots (these equations are commonly called the *reduced equations*) are found to be of a degree less than the degree of the proposed equation, or are at least decomposable into other equations of degree less than it; 2° that one can easily deduce from them the values of the roots one is searching for.

The art of solving equations consists of discovering functions of the roots, which have the properties which we stated. But is it always possible to find such functions for equations of any degree, that is, for the desired number of roots? That is certainly very difficult to judge in general.

With regard to equations whose degree does not exceed four, the simplest functions which give their solution can be represented by the general formula

$$x' + yx'' + y^2x''' + \cdots + y^{\mu-1}x^{(\mu)},$$

where $x', x,'' x',''' \ldots, x^{(\mu)}$ are the roots of the proposed equation, which one supposes to be of degree μ, and y is a root of the equation

$$y^\mu - 1 = 0,$$

which is different from 1, that is, a root of the equation

$$y^{\mu-1} + y^{\mu-2} + y^{\mu-3} + \cdots + 1 = 0.$$

This is a consequence of everything we have shown in the first two sections in relation to the solution of equations of degree three and four.

Even though Lagrange ends his memoir on a rather pessimistic note, the work represents a major accomplishment that paved the way for later progress. Galois picked up directly where Lagrange left off. In the first part of the excerpt in this section Lagrange makes a detailed analysis of the *reduced equation* associated to the *proposed equation*, which is to be solved. He shows that for equations of degree three and four the roots of the reduced equation can be written as a function of the roots of the proposed equation, with roots of unity as coefficients. In both cases the reduced equation can be solved like an equation of degree lower than that of the proposed equation, even though its degree is higher. And, of course, the roots of the proposed equation are rational functions of the roots of the reduced equation.

He then concludes that the key to solving higher-degree equations is to find such higher degree analogues of the reduced equation, whose roots are functions of the roots of the proposed equation. In addition, of course, the reduced equation should be solvable. The reduced equation is now commonly called a *Lagrange resolvent* of the proposed equation. A large part of his memoir is devoted to studying the properties of such resolvents, if they indeed exist. Lagrange's final conclusion is that while it has not been shown that this method is impossible, it is "difficult to judge" whether there is anything beyond luck that one can rely on.

Not long afterwards, Gauss carried out exactly Lagrange's program for the cyclotomic polynomials, as remarked in the Introduction. One very strong requirement Lagrange makes is that the reduced equation should be solvable, which is necessary in order to actually solve the proposed equation. In the next section we will see that if this requirement is dropped, then resolvents always exist.

Exercise 5.14: Solve the equation $x^3 - 1 = 0$ by Cardano's method.

Exercise 5.15: Solve the equation $x^3 = 15x + 4$ by the method of Cardano, as explained by Lagrange. Express the roots of $x^3 = 15x + 4$ in terms of the roots of the reduced equation.

Exercise 5.16: Make a summary of the source in this section. Contrast Lagrange's investigations with Cardano's work.

5.5   Galois Ends the Story

The story of Evariste Galois is one of the gems of mathematical history. His life is summed up admirably by H. Edwards in his book on Galois theory [48]:

> Great mathematicians usually have undramatic lives, or, more precisely, the drama of their lives lies in their mathematics and cannot be appreciated by nonmathematicians. The great exception to this rule is Evariste Galois (1811–1832). Galois' life story—what we know of it—is like a romantic novel. In secondary school, he was denied admission to the Ecole Polytechnique, which was the premier institution of higher learning in mathematics at the time, and the mathematical establishment ignored, mislaid, lost, and failed to understand his treatises. Meanwhile, he was persecuted for his political activities and spent many months in jail as a political prisoner. At the age of 20 he was killed in a duel involving, in some mysterious way, honor and a woman. On the eve of the fatal duel he wrote a letter to a friend

PHOTO 5.10. Galois.

outlining his mathematical accomplishments and asking that the friend try
to bring his work to the attention of the mathematical world. Against great
odds, Galois' few supporters did finally, 14 years after his death, succeed
in finding an audience for his work, and portions of his writings were published in 1846 by Joseph Liouville in his *Journal de Mathematiques*. After
that, recognition of the great importance of his work came very quickly,
and Galois began to be regarded, as he is today, as one of the great creative
mathematicians of all time [48, p. 1].

Needless to say, Galois was not very favorably disposed toward the French
scientific establishment. The following excerpt is from an unpublished preface to
one of his memoirs.

First of all, the second page of this work is not encumbered with names,
forenames, qualities, titles, and elegies of some miserly prince whose purse
will be opened with the fumes of incense—with the threat of being closed
when the censer-bearer is empty. You will not see, in characters three times
larger than the text, a respectful homage to someone of high position in the
sciences, to a wise protector—something indispensable (I was going to say
inevitable) to someone twenty years old who wants to write. I do not say
to anyone that I owe to their advice and their encouragement everything

that is good in my work. I do not say it because that would be to lie. If
I had to address anything to the great in the world or the great in science
(at the present time the distinction between these two classes of people is
imperceptible) I swear that it would not be in thanks. I owe to the ones
that I have published the first of these two memoirs so late, to the others
that I have written it all in prison, a place it would be wrong to consider a
place of meditation, and where I am often amazed at my self-restraint and
keeping my mouth shut in the face of my stupid ill-natured critics [Zoiles];
I think I may use the word Zoiles without fear of being immodest when
my adversaries are in my mind base. It is not my subject to say how and
why I was detained in prison, but I must say how my manuscripts have been
lost most often in the cartons of Messieurs the members of the Institute,
although in truth I cannot imagine such thoughtlessness on the part of men
who have the death of Abel on their consciences. For me, who does not want
to be compared with that illustrious mathematician, it suffices to say that
my memoir on the theory of equations was deposited, in substance, with the
Academy of Sciences in February 1830, that extracts from it had been sent
in 1829, that no report on it was then issued and it has been impossible to
recover the manuscript.... [58, p. 504]

Biographical information about Galois's life may be found in [42, 46, 94, 143].

We now come to Galois's main paper on the theory of equations. It represented
a formidable challenge to even his most astute mathematical contemporaries. A
proper mathematical analysis requires a considerable background in abstract algebra and is carried out in [48]. The English translation of the memoir that we
use here is contained in an appendix to [48]. We shall limit ourselves to a small
excerpt in order to get a sense of the flavor of Galois's mathematics and how it differs dramatically from that of his predecessors. It is beyond the scope of this book
and its intended audience to provide sufficient annotation for a full understanding
and appreciation of the finer points of Galois's tremendous accomplishment. Thus,
the reader unfamiliar with abstract algebra may better appreciate the situation that
Galois's contemporaries found themselves in when confronted with his paper.
However, the following brief introduction should convey the gist of the entirely
new paradigm that Galois proposed for the subject. We follow closely the synopsis
given in Appendix 2 of [48]. The reader interested in a more detailed exposition of
Galois's work from a modern point of view, as well as a more in-depth discussion
of Galois's paper, is encouraged to study the wonderful exposition in [48].

As with Lagrange, the basic object under investigation is an equation

$$f(x) = 0,$$

where the left side $f(x)$ is a polynomial whose coefficients are to be thought of
as unspecified numbers. Galois first defines what he means by an equation being
reducible, as opposed to being irreducible. Of course, he means that an equation
is reducible if the polynomial $f(x)$ factors. But whether a polynomial factors or
not depends very much on which set of numbers one considers as "allowable"

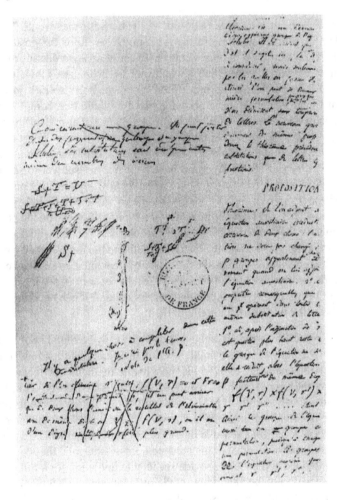

PHOTO 5.11. Page from Galois's papers.

in the factorization. Over the complex numbers **C** every polynomial factors into a product of linear factors, according to the Fundamental Theorem of Algebra, but if we allow only real numbers, then the best we can do is factorization into linear and quadratic polynomials; many quadratic ones are irreducible "over **R**." Likewise, the equation $x^2 - 2 = 0$ is irreducible when considered over the rational numbers **Q**, but becomes reducible if we consider as allowable, or "known," in Galois's terminology, certain irrational numbers such as $\sqrt{2}$. In order to solve the problem at hand, namely to find a formula for the roots of $f(x) = 0$, it is of course sufficient to assume that $f(x)$ is irreducible, which turns out to be a useful assumption. Thus, Galois concludes that it is important to specify not just the equation under consideration, but also the "field" of numbers that one considers as known, and over which the equation is irreducible. If at some point we enlarge our field of numbers over which we are considering the equation, then we are *adjoining* these new numbers to the equation, in Galois's terminology. In order to understand Galois, we need to be a bit more specific about what kinds of fields of numbers we are considering, and what the numbers in the new field have to do with the old ones.

The key concept in what follows is that of a subfield of the complex numbers **C**. First of all, the set of rational numbers **Q**, the real numbers **R**, and **C** are all examples of such fields. Briefly, a field K in our situation is a set of complex numbers that is closed under addition and multiplication, so that for every number in K its negative is in K, and for every nonzero number in K its reciprocal is in K. Besides the ones mentioned, other relevant examples are the following (in each case, the reader is encouraged to verify that the set indeed has the required properties).

1. Let $K = \{a + b\sqrt{2} \mid a, b \in \mathbf{Q}\}$, which we denote by $\mathbf{Q}(\sqrt{2})$. It has the property that it contains all the roots of the equation $x^2 - 2 = 0$; that is, the polynomial $x^2 - 2$, which is irreducible over **Q**, decomposes into linear factors over the larger field K,

$$x^2 - 2 = (x - \sqrt{2})(x + \sqrt{2}).$$

It has the further property that it is the smallest subfield of the complex numbers that contains **Q** and $\sqrt{2}$. Thus, if we adjoin $\sqrt{2}$ to our equation it becomes reducible. In fact, even better, the new field contains *all* the roots of the equation. Such a field is now commonly called a *splitting field* of the polynomial, since over this field, it splits into a product of linear factors.

2. Next consider the equation $x^3 - 2 = 0$, the left side of which is an irreducible polynomial, if we consider it over **Q**. If we proceed as in the previous example, then we obtain the field

$$K = \{a + b\sqrt[3]{2} + c\sqrt[3]{2^2} \mid a, b, c \in \mathbf{Q}\} = \mathbf{Q}(\sqrt[3]{2}).$$

If we now adjoin $\sqrt[3]{2}$ to the equation, then we can factor it as

$$x^3 - 2 = (x - \sqrt[3]{2})(x^2 + \sqrt[3]{2}x + \sqrt[3]{2^2}).$$

Using the quadratic formula, the reader may check immediately that the quadratic factor has imaginary roots and is therefore irreducible over the field $\mathbf{Q}(\sqrt[3]{2})$, which contains only real numbers. Since its roots are $\zeta\sqrt[3]{2}$ and $\zeta^2\sqrt[3]{2}$, where ζ is a primitive cube root of unity (that is, $\zeta^2 \neq 1$), we see that a splitting field for the equation $x^3 - 2 = 0$ is the field

$$\mathbf{Q}(\sqrt[3]{2}, \zeta) = \{a + b\sqrt[3]{2} + c\sqrt[3]{2^2} + d\zeta + e\zeta\sqrt[3]{2} + f\zeta\sqrt[3]{2^2} \mid a, b, c, d, e, f \in \mathbf{Q}\}.$$

Recall that ζ is a root of the equation

$$x^2 + x + 1 = 0.$$

Thus, the field

$$\mathbf{Q}(\zeta) = \{a + b\zeta \mid a, b \in \mathbf{Q}\}$$

is a splitting field for the equation $x^3 - 1 = 0$.

In general, if $f(x) = 0$ is a polynomial equation with coefficients in a field K (which may be larger than $\mathbf{Q}$), and the roots of $f(x)$ are $a, b, c, \ldots$, then a splitting field for $f(x)$ is the smallest subfield of $\mathbf{C}$ that contains all the roots $a, b, c, \ldots$. We shall denote such a field by $K(a, b, c, \ldots)$. One can show that every element of $K(a, b, c, \ldots)$ is a sum of the roots $a, b, c, \ldots$ with coefficients in K.

3. It was proven by Carl Louis F. Lindemann (1852–1939) in 1882 that π is a transcendental number, which means that there is no algebraic equation with rational numbers as coefficients that has π as a root. (See the set theory chapter.) Then what does the field $\mathbf{Q}(\pi)$ look like, the smallest subfield of $\mathbf{C}$ that contains $\mathbf{Q}$ and the number π? Since it is closed under addition and multiplication, it certainly has to contain all powers of π, all products of powers of π with rational numbers, as well as all sums of such products. Another way of saying this is that $\mathbf{Q}(\pi)$ has to contain all polynomials with rational coefficients evaluated at π. Since the field also has to be closed under multiplicative inverses, it also contains all the reciprocals of nonzero polynomials. Since π does not satisfy an algebraic equation with rational coefficients, this amounts to simply renaming the variable of the polynomials to be π. Thus, the field $\mathbf{Q}(\pi)$ can be thought of as consisting of all rational functions in one variable, which we may call t (or π), with rational coefficients. In proper terminology, $\mathbf{Q}(\pi) = \mathbf{Q}(t)$ is a *field of rational functions* over $\mathbf{Q}$.

One of the first observations one makes in a modern treatment of Galois theory is that every algebraic equation $f(x) = 0$ has a splitting field, something that Galois seems to have taken for granted. The first order of business in Galois's memoir is to give an explicit description of this splitting field. So, let $f(x) = 0$ be a polynomial equation with roots $a, b, c, \ldots$, the proposed equation, in Lagrange's terminology. We will assume that the field from which the coefficients of $f(x)$ are taken is some subfield K of $\mathbf{C}$, which may or may not be equal to $\mathbf{Q}$. Let $L = K(a, b, c, \ldots)$ be the splitting field of $f(x)$, another subfield of $\mathbf{C}$, which will include K, but will in general be larger than K. For the sake of simplicity, the reader may just assume that $K = \mathbf{Q}$. Let the degree of $f(x)$ be n, and consider a linear polynomial

$$T = AU + BV + CW + \cdots$$

in n variables $U, V, W, \ldots$ with integer coefficients $A, B, C, \ldots$. In the splitting field L, the polynomial $f(x)$ has n roots, and there are $n!$ ways of substituting those roots for the variables $U, V, W, \ldots$. For each such substitution we obtain an element of L. In the case that these $n!$ elements are all *distinct*, we call T a *Galois resolvent* of $f(x) = 0$ over K. It can be shown [48, §32] that if the roots of $f(x) = 0$ are all distinct, then one can make a choice of coefficients $A, B, C, \ldots$ such that T takes on $n!$ distinct values, hence is a Galois resolvent for the equation.

Galois proves that if t is one of these $n!$ distinct values in L, then the field L is of the form $K(t)$ (see the above examples), so that every root of $f(x) = 0$ can be expressed as a rational function of t, that is, as the quotient of two polynomials in t with coefficients in K. Now suppose that $t_1, t_2, \ldots, t_{n!}$ are all the distinct elements.

Then one can show that after multiplying out the product

$$F(X) = (X - t_1)(X - t_2) \cdots (X - t_{n!}),$$

one obtains a polynomial in X with coefficients that actually lie in K. Now decompose $F(X)$ into factors that are irreducible over K, $F(X) = G_1(X) \cdots G_s(X)$. The roots of each G_i are a subset of the roots $t_1, \ldots, t_{n!}$ of $F(X)$, and Galois showed that if t is any root of any of the G_i, then the splitting field L for $f(x)$ is equal to $K(t)$. In this way one obtains a polynomial that is irreducible over K, and any root of this polynomial generates $K(t)$. This result provides a very explicit description of L.

What has Galois accomplished now, in light of the program that Lagrange laid out? Given a proposed equation, Lagrange wanted to find an associated reduced equation whose roots could be expressed as linear polynomials in the roots of the proposed equation, with roots of unity as coefficients, so that in turn, the roots of the proposed equation were rational functions of the roots of this reduced equation. Galois first produces a candidate t for the root of an as yet undetermined reduced equation, namely the value of a Galois resolvent T under any substitution of the roots of $f(x)$. Since t is an element of the splitting field $L = K(a, b, c, \ldots)$, it can be written as a linear combination of the roots $a, b, c, \ldots$, with coefficients in K. Conversely, since $L = K(t)$, one can write the roots $a, b, c, \ldots$ as rational functions of t. Now all that is needed is a polynomial with coefficients in K that has t as a root. That is then the candidate for producing the reduced equation that Lagrange was after. And of course, that is given by any of the G_i above. Thus, Galois has shown that this part of Lagrange's program can be carried out in principle, even though he does not indicate a procedure that one could use for a specific polynomial $f(x)$.

Galois now associates to the equation $f(x) = 0$ a group of permutations of the roots $a, b, c, \ldots$ as follows. (In his paper Galois distinguishes permutations, arrangements of a given set of objects, from substitutions, the act of rearranging these objects. This distinction is not customary anymore.) Let $G(X)$ be an irreducible factor of degree r of the polynomial $F(X)$ above, and let $t_1, t_2, \ldots, t_r$ be all of its roots. (One can show that if the coefficients of $G(X)$ lie in a subfield of $\mathbf{C}$, then $G(X)$ has r distinct roots.) Then

$$L = K(a, b, c, \ldots) = K(t_i),$$

for any choice of a root t_i. Suppose that $f(x)$ has n distinct roots, which we will label $a_1, \ldots, a_n$. Then each root can be expressed as a rational function of t, that is, there exist rational functions $h_1(X), \ldots, h_n(X)$ such that

$$a_i = h_i(t),$$

for $i = 1, \ldots, n$. The same holds true if we replace t by any other of the t_j. Therefore, for each $j = 1, \ldots, r$, the elements

$$h_1(t_j), h_2(t_j), \ldots, h_n(t_j)$$

are the roots $a_1, a_2, \ldots, a_n$ of $f(x)$ in some order. Therefore, we obtain a permutation σ_j of the roots of $f(x)$ by setting $\sigma_j(a_1) = h_1(t_j)$, $\sigma_j(a_2) = h_2(t_j), \ldots$, for every $j = 1, \ldots, r$. The *Galois group* of $f(x) = 0$ over K is the group (under composition) of the r permutations obtained in this way. (The reader unfamiliar with the basics of group theory may safely substitute the word "set" for "group.") One now needs to show, of course, that this Galois group comes out to be the same regardless of what choices one makes for the irreducible factor G and its root t, all of which can be done. If the field K is extended to a larger field K', and one views the equation $f(x) = 0$ as defined over K', then the Galois group of the equation over K' will be a subgroup of the Galois group over K.

Herein lies one of the great accomplishments of Galois. Lagrange pointed the way toward the search for "resolvents," whose roots determine the roots of the given equation via rational functions. But Lagrange added the extra condition that the resolvent ought to be itself solvable. By dropping this requirement Galois could show that such a resolvent G always exists (at least for equations without multiple roots). His proof was even constructive enough to give a procedure for finding one.

His second great accomplishment was to give a criterion for when an equation was solvable by radicals. This condition applied to the Galois group of the equation, and is now known as "solvability" of the group. Thus, an equation is solvable by radicals if and only if its Galois group is solvable. In this language, Abel had already shown that an equation was solvable by radicals if its Galois group was commutative, which means that applying two permutations in either order gives the same result. Such groups are now knows as "abelian," in honor of Abel's accomplishments.

Generally speaking, the criterion works as follows. Suppose we are given an equation $f(x) = 0$, and let us assume that its coefficients are rational numbers. Is this equation solvable by radicals? According to Galois, we need to look at the Galois group of $f(x)$. The answer is yes, if it is possible to extend the field $\mathbf{Q}$ of known quantities successively, by adjoining radicals, that is, roots of polynomials of the form $X^m - a$ for suitable values of a in the field of known quantities, so that ultimately the Galois group of $f(x)$ over the enlarged field of known quantities consists of the identity permutation alone. So the question of which equations are solvable by radicals is equivalent to the question of which reductions of Galois groups can be accomplished by adjunctions of this type. Galois then goes on to give a solution of this group-theoretic problem.

Of course, it is not the kind of solution Lagrange was looking for. A new era in the theory of polynomial equations had begun. Nor was Galois's solution a really practical one, since the calculations involved are immense, in general. He, of course, realized that himself.

> If you now give me an equation that you have chosen at your pleasure, and if you want to know if it is or is not solvable by radicals, I could do nothing more than to indicate to you the means of answering your question, without wanting to give myself or anyone the task of doing it. In a word, the calculations are impractical.

From that, it would seem that there is no fruit to derive from the solution that we propose. Indeed, it would be so if the question arose from this point of view. But, most of the time, in the applications of Algebraic Analysis, one is led to equations of which one knows beforehand all the properties: properties by means of which it will always be easy to answer the question by the rules we are going to explain.... All that makes this theory beautiful and at the same time difficult, is that one has always to indicate the course of analysis and to foresee its results without ever being able to perform [the calculations] [169, pp. 313–314].

Evariste Galois, from
Memoir on the Conditions for Solvability of Equations by Radicals

PRINCIPLES

I shall begin by establishing some definitions and a sequence of lemmas, all of which are known.

Definitions. An equation is said to be reducible if it admits rational divisors; otherwise it is irreducible.

It is necessary to explain what is meant by the word rational, because it will appear frequently.

When the equation has coefficients that are all numeric and rational, this means simply that the equation can be decomposed into factors which have coefficients that are numeric and rational.

But when the coefficients of an equation are not *all* numeric and rational, one must mean by a rational divisor a divisor whose coefficients can be expressed as rational functions of the coefficients of the proposed equation, and, more generally, by a rational quantity a quantity that can be expressed as a rational function of the coefficients of the proposed equation.

More than this: one can agree to regard as rational all rational functions of a certain number of determined quantities, supposed to be known *a priori*. For example, one can choose a particular root of a whole number and regard as rational every rational function of this radical.

When we agree to regard certain quantities as known in this manner, we shall say that we *adjoin* them to the equation to be resolved. We shall say that these quantities are *adjoined* to the equation.

With these conventions, we shall call *rational* any quantity which can be expressed as a rational function of the coefficients of the equation and of a certain number of *adjoined* quantities arbitrarily agreed upon.

When we make use of auxiliary equations, they will be rational if their coefficients are rational in our sense.

One sees, moreover, that the properties and the difficulties of an equation can be altogether different, depending on what quantities are adjoined to it. For example, the adjunction of a quantity can render an irreducible equation reducible.

Thus, when one adjoins to the equation

$$\frac{x^n - 1}{x - 1} = 0, \text{ where } n \text{ is prime,}$$

a root of one of Mr. Gauss's auxiliary equations, this equation decomposes into factors, and consequently becomes reducible.

$$\cdots$$

Lemma II. Given any equation with distinct roots $a, b, c, \ldots$, one can always form a function V of the roots such that no two of the values one obtains by permuting the roots in this function are equal.

For example, one can take

$$V = Aa + Bb + Cc + \cdots,$$

$A, B, C, \ldots$ being suitably chosen whole numbers.

Lemma III. When the function V is chosen as indicated above, it will have the property that all the roots of the given equation can be expressed as rational functions of V.

$$\cdots$$

PROPOSITION I

Theorem. Let an equation be given whose m roots are $a, b, c, \ldots$. There will always be a group of permutations of the letters $a, b, c, \ldots$ which will have the following property:

1. that each function invariant[8] under the substitutions of this group will be known rationally;
2. conversely, that every function of the roots which can be determined rationally will be invariant under these substitutions.

(In the case of algebraic equations, this group is none other than the set of all $1 \cdot 2 \cdot 3 \cdots m$ permutations of the m letters, because in this case the symmetric functions are the only ones that can be determined rationally.)

---

[8] Here we call a function invariant not only if its form is unchanged by the substitutions of the roots, but also if its numerical value does not vary when these substitutions are applied. For example, if $Fx = 0$ is an equation, Fx is a function of the roots which is not changed by any substitution.

When we say that a function is rationally known, we mean that its numerical value can be expressed as a rational function of the coefficients of the equation and the quantities that have been adjoined. (Translator's note.)

(In the case of the equation $(x^n - 1)/(x - 1) = 0$, if one supposes that $a = r$, $b = r^g, c = r^{g^2}, ...,$ g being a primitive root, the group of permutations will be simply this one:

$$abcd ... k,$$
$$bcd ... ka,$$
$$cd ... kab,$$
$$.................$$
$$kabc ... i \qquad [\text{sic; } i \text{ precedes } k].$$

In this particular case, the number of permutations is equal to the degree of the equation, and the same will be true for equations all of whose roots are rational functions of one another.)

. . .

We will call the group in question the group of the equation.

...)

Galois gives two parenthetic examples for the Galois group of an equation, both rather cryptic. The first example of "algebraic equations" refers to the case where the field of known quantities is obtained as follows. Let K be a field, and construct the field K' by adjoining n indeterminates $Y_1, \ldots, Y_n$ to K, that is, $K' = K(Y_1, \ldots, Y_n)$. Then Galois claims that the Galois group of the equation

$$x^n + Y_1 x^{n-1} + Y_2 x^{n-2} + \cdots + Y_n = 0$$

is the full group of all permutations of n letters, which contains $n!$ elements. For a complete proof of this assertion see [48, p. 89]. For a discussion of the second example see [48, pp. 93–94].

PROPOSITION V

Problem. In which case is an equation solvable by simple radicals?

I shall observe first that in order to solve an equation it is necessary to reduce its group successively until it contains only one permutation. For, when an equation is solved, any function whatever of its roots is known, even when it is not invariant under any permutation.

With this set forth, let us try to find the condition which the group of an equation should satisfy in order that it can be thus reduced by the adjunction of radical quantities.

Let us follow the sequence of possible operations in this solution, considering as distinct operations the extraction of each root of prime degree.

Adjoin to the equation the first radical to be extracted in the solution. One of two things can happen: either by the adjunction of this radical the group of permutations of the equation will be diminished, or, this extraction of a root being only a preparation, the group will remain the same.

In any case, after a certain *finite* number of extractions of roots the group must find itself diminished because otherwise the equation would not be solvable.

If at this point it occurs that there are several ways to diminish the group of the given equation by the simple extraction of a root, it is necessary, in what we are going to say, to consider only a radical of the least possible degree among all the simple radicals which are such that the knowledge of each of them diminishes the group of the equation.

Exercise 5.17: Summarize the above part of Galois's memoir and contrast it to Lagrange's work. What are the main differences in their approach and in the results they obtain?

Exercise 5.18: Show that $x^3 - 2$ is irreducible over the field $\mathbf{Q}$ of rational numbers.

Exercise 5.19: Solve $x^3 - 2 = 0$ by Cardano's method, as described by Lagrange.

Exercise 5.20: Show that $T = U - V$ is a Galois resolvent for $x^3 - 2 = 0$.

Exercise 5.21: Let t be obtained from T by substituting any two roots of $x^3 - 2$ into T. Find a sixth-degree polynomial over $\mathbf{Q}$ of which t is a root. (Hint: Use the method explained in the introduction to Galois's memoir.) Find a condition for this polynomial to factor. In this case find its factorization. (It factors as a product of two irreducible cubic polynomials.)

Exercise 5.22: Let a, b, c be the roots of $x^3 - 2 = 0$ and let $t = a - b$. Verify that

$$a = \frac{t^3 + 6}{2t^2}.$$

(The other roots are also rational functions of t. See [48].)

Exercise 5.23: Find the Galois group of $x^3 - 2$.

References

[1] N.H. Abel, *Oeuvres Complètes de N.H. Abel, mathématicien, Nouvelle édition*, M.M.L. Sylow and S. Lie (eds.), 2 vols., Oslo, 1881. (See also Johnson Reprint Corp., 1964.)

[2] L.M. Adleman and D.R. Heath-Brown, *The First Case of Fermat's Last Theorem*, Invent. Math. **79** (1985), 409–416.

[3] Archimedes, *Works*, T.L. Heath (ed.), Dover, New York.

[4] Aristotle, *Physics, vol. II*, P. Wicksteed and F. Cornford (transl.), Harvard University Press, Cambridge, 1960.

[5] W. Aspray and P. Kitcher, *History and Philosophy of Modern Mathematics*, University of Minnesota Press, Minneapolis, MN, 1988.

[6] A. Baker, *A Concise Introduction to the Theory of Numbers*, Cambridge University Press, New York, 1984.

[7] K. Barner, *Paul Wolfskehl and the Wolfskehl Prize*, Notices of the Amer. Math. Soc. **44** (1997), 1294–1303.

[8] M. Baron, *The Origins of the Infinitesimal Calculus*, Dover, New York, 1969.

[9] I. Bashmakova, *Diophantus and Diophantine Equations*, Math. Assoc. of Amer., Washington, D.C., 1997.

[10] P. Beckmann, *A History of Pi*, Dorset Press, New York, 1971.

[11] P. Benacerraf and H. Putnam, *Philosophy of Mathematics*, Cambridge University Press, Cambridge, 2nd ed., 1983; transl. from Mathematische Annalen (Berlin) vol. 95 (1926), 161–190.

[12] J.L. Berggren, *Episodes in the Mathematics of Medieval Islam*, Springer-Verlag, New York, 1986.

[13] N. Biggs, E. Lloyd, and R. Wilson, *Graph theory 1736–1936*, Oxford University Press, Oxford, 1976.

[14] G. Birkhoff and U. Merzbach, *A Source Book in Classical Analysis*, Harvard University Press, Cambridge, Mass., 1973.

[15] B. Bolzano, *Theory of Science* (transl. by R. George), University of California Press, Berkeley, 1972.

[16] B. Bolzano, *Paradoxes of the Infinite*, Yale University Press, New Haven, 1950, translation of *Paradoxien des Unendlichen*, Leipzig, 1851.

[17] B. Bolzano, *Paradoxien des Unendlichen*, Mayer & Müller, Berlin, 1889.

[18] R. Bonola, *Non-Euclidean Geometry*, 1906, Dover, New York, 1955.

[19] U. Bottazzini, *The Higher Calculus: A History of Real and Complex Analysis from Euler to Weierstrass*, Springer, New York, 1986.

[20] C. Boyer, *A History of Mathematics*, John Wiley & Sons, New York, 1968.

[21] C. Boyer, *The History of the Calculus and its Conceptual Development*, Dover, New York, 1949.

[22] C. Boyer, *History of Analytic Geometry*, The Scholar's Bookshelf, Princeton, 1988.

[23] L.L. Bucciarelli and N. Dworsky, *Sophie Germain: An Essay in the History of the Theory of Elasticity*, D. Reidel Publishing Co., Dordrecht, Holland, 1980.

[24] R. Calinger (ed.), *Vita Mathematica: Historical Research and Integration with Teaching*, Math. Assoc. of Amer., Washington, D.C., 1996.

[25] R. Calinger (ed.), *Classics of Mathematics*, 2nd edition, Prentice-Hall, Englewood Cliffs, New Jersey, 1995.

[26] G. Cantor, *Contributions to the Founding of the Theory of Transfinite Numbers*, Dover, New York, 1952.

[27] *Briefwechsel Cantor–Dedekind*, E. Noether and J. Cavaillès (eds.), 1937, Hermann, Paris.

[28] G. Cantor, *Gesammelte Abhandlungen*, E. Zermelo (ed.), Georg Olms Verlagsbuchhandlung, Hildesheim, 1962.

[29] G. Cardano, *The Great Art or The Rules of Algebra*, transl. into English by T.R. Witmer, MIT Press, Cambridge, 1968; from the original *Ars Magna*, Nürnberg, 1545.

[30] A. Cauchy, *Oeuvres Complètes (2)*, Académie des Sciences, 1882–1981.

[31] A. Cauchy, *Résumé ... sur le Calcul Infinitésimal*, Edition Marketing, Paris, 1994.

[32] B. Cavalieri, *Exercitationes Geometricae Sex*, Bologna, 1647, reprinted by Istituto Statale d'Arte di Urbino, 1980.

[33] D. Cox, *Introduction to Fermat's Last Theorem*, American Mathematical Monthly **101** (1994), 3–14.

[34] H.S.M. Coxeter, *Introduction to Geometry*, Wiley, New York, 1969.

[35] A.D. Dalmédico, *Sophie Germain*, Scientific American, December 1991, 77–81.

[36] J. Dauben, *Georg Cantor: His Mathematics and Philosophy of the Infinite*, Princeton University Press, Princeton, NJ, 1990.

[37] J. Dauben, *Abraham Robinson: The Creation of Nonstandard Analysis: A Personal and Mathematical Odyssey*, Princeton University Press, Princeton, NJ, 1995.

[38] R. Dedekind, *Essays on the Theory of Numbers*, Dover, New York, 1963.

[39] Diophantus, *Diophanti Alexandrini arithmeticorum libri sex, et de numeris multangulis liber unus*, Toulouse, 1670.

[40] L.E. Dickson, *Introduction to the Theory of Numbers*, University of Chicago Press, Chicago, 1929.

[41] L.E. Dickson, *History of the Theory of Numbers*, Chelsea Publishing, New York, 1992.

[42] *Dictionary of Scientific Biography*, C.C. Gillispie and F.L. Holmes (eds.), Scribner, New York, 1970.

[43] E.J. Dijksterhuis, *Archimedes*, Princeton University Press, Princeton, New Jersey, 1987.

[44] W. Dunham, *A "Great Theorems" Course in Mathematics*, Amer. Math. Monthly **93** (1986), 808–811.

[45] W. Dunham, *Journey Through Genius*, John Wiley, New York, 1980.

[46] P. Dupuy, *La Vie d'Evariste Galois*, Annales de l'Ecole Normale, vol. III (1896), 197–266.

[47] H.M. Edwards, *Fermat's Last Theorem: A Genetic Introduction to Number Theory*, Springer Verlag, New York, 1977.

[48] H.M. Edwards, *Galois Theory*, Springer Verlag, New York, 1984.

[49] *Encyclopedia Britannica*, 15th ed., 1986.

[50] F. Engel and P. Stäckel, *Die Theorie der Parallellinien von Euklid bis auf Gauss*, Teubner, Leibzig, 1895.

[51] Euclid, *The Thirteen Books of Euclid's Elements*, T.L. Heath (ed.), Dover, New York, 1956.

[52] L. Euler, "Theorematum quorundam arithmeticorum demonstrationes," *Commentarii Academiae Scientiarum Petropolitanae* 10, 1738 publ. 1747, pp. 125–46, in *Opera omnia*, ser. I, vol. 2, pp. 36–58.

[53] L. Euler, *Vollständige Anleitung zur Algebra*, Saint Petersburg, 1770, in *Opera omnia*, ser. I, vol. 1 (see also ser. I, vol. 5) (English translation: *Elements of Algebra*, translated by John Hewlett, Springer-Verlag, New York, 1984).

[54] L. Euler, *Correspondence Mathématique et Physique de Quelques Célèbres Géomètres du XVIIIème Siècle*, Tome I, P.-H. Fuss (ed.), St. Petersburg, 1843, reprinted in *The Sources of Science*, No. 35, Johnson Reprint Co., New York and London, 1968.

[55] L. Euler, *Introduction to analysis of the infinite* (transl. John D. Blanton), Springer-Verlag, New York, 1988.

[56] H. Eves, *Great Moments in Mathematics (before 1650)*, Mathematical Association of America, Washington, DC, 1983.

[57] A. Fantoli (transl. G.V. Coyne), *Galileo*, Studi Galileiani, vol. 3, Vatican Observatory, 1994, 480–488.

[58] J. Fauvel and J. Gray (eds.), *The History of Mathematics: A Reader*, MacMillan Press, London, 1987.

[59] P. de Fermat, *Oeuvres*, P. Tannery (ed.), Paris, 1891–1922.

[60] M. Fierz, *Girolamo Cardano*, Birkhäuser, Boston, 1983.

[61] A. Fraenkel, *On the Foundations of Cantor–Zermelo Set Theory*, Mathematische Annalen **86** (1922), 230–237.

[62] A. Fraenkel and Y. Bar-Hillel, *Foundations of Set Theory*, North-Holland, Amsterdam, 1958.

[63] A. Fraenkel, *Abstract Set Theory*, North-Holland, Amsterdam, 1953.

[64] J. Friberg, *Methods and Traditions of Babylonian Mathematics*, Historia Mathematica **8** (1981), 277–318.

[65] G. Galilei, *Dialogues Concerning Two New Sciences*, Dover, New York, 1954.

[66] E. Galois, *Ecrits et Mémoires Mathématiques d'Evariste Galois*, (R. Bourgne and J.-P. Azra, eds.) Gauthiers-Villars, Paris, 1962.

[67] L. Gårding and C. Skau, *Niels Henrik Abel and Solvable Equations*, Arch. Hist. Exact Sci. **48** (1994), 81–103.

[68] C.F. Gauss, *Disquisitiones Arithmeticae* (English translation), Yale University Press, New Haven, 1966.

[69] S. Germain, *Unpublished letter to C.F. Gauss*, Niedersächsische Staats- und Universitätsbibliothek Göttingen.

[70] S. Germain, *Papiers de Sophie Germain*, MS. FR9114, Bibliothèque Nationale, Paris.

[71] P. Giblin, *Primes and Programming*, Cambridge University Press, New York, 1993.

[72] J. Grabiner, *The Origins of Cauchy's Rigorous Calculus*, M.I.T. Press, Cambridge, Mass., 1981.

[73] I. Grattan-Guinness, *Towards a Biography of Georg Cantor*, Annals of Science **27** (1971), 345–391.

[74] I. Grattan-Guinness, *An Unpublished Paper by Georg Cantor: Principien einer Theorie der Ordnungstypen. Erster Mittheilung*, Acta Math. **124** (1970), 65–107.

[75] I. Grattan-Guinness, *The Development of the Foundations of Mathematical Analysis from Euler to Riemann*, M.I.T. Press, Cambridge, Mass., 1970.

[76] I. Grattan-Guinness, *The Rediscovery of the Cantor–Dedekind Correspondence*, Jahresbericht d. Deutsch. Math. Verein. **76** (1974), 104–139.

[77] I. Grattan-Guinness (ed.), *From the Calculus to Set Theory, 1630–1910: An Introductory History*, Gerald Duckworth and Co., London, 1980.

[78] J. Gray, *Ideas of Space: Euclidean, Non-Euclidean, and Relativistic*, 2nd ed., Oxford University Press, 1989.

[79] L. Grinstein and P. Campbell (eds.), *Women of Mathematics: A Biobibliographic Sourcebook*, Greenwood Press, New York, 1987.

[80] E. Grosswald, *Representations of Integers as Sums of Squares*, Springer-Verlag, New York, 1985.

[81] E. Hairer, G. Wanner, *Analysis by its History*, New York, Springer-Verlag, 1996.

[82] R.R. Hamburg, *The Theory of Equations in the 18th Century: The Work of Joseph Lagrange*, Archive for the History of the Exact Sciences **16** (1976/7), 17–36.

[83] T. Hawkins, *Lebesgue's Theory of Integration: Its Origins and Development*, Chelsea, New York, 1975.

[84] T.L. Heath, *A History of Greek Mathematics*, Vols. I and II, Oxford University Press, London, 1921.

[85] T.L. Heath, *A Manual of Greek Mathematics*, Dover, New York, 1963.

[86] T.L. Heath, *Diophantus of Alexandria*, Dover, New York, 1964.

[87] J. van Heijenoort, *From Frege to Gödel, A Source Book in Mathematical Logic, 1879–1931*, Harvard University Press, Cambridge, 1967.

[88] D. Hilbert, *The Foundations of Geometry*, translated by E.J. Townsend, Open Court, LaSalle, Illinois, 1902.

[89] D.R. Hofstadter, *Gödel, Escher, Bach: an Eternal Golden Braid*, Basic Books Inc., New York, 1979.

[90] *International Study Group on the History and Pedagogy of Mathematics Newsletter*.

[91] G. Joseph, *The Crest of the Peacock: Non-European Roots of Mathematics*, Penguin Books, London, 1992.

[92] J.-P. Kahane and P. Lemarié-Rieusset, *Fourier Series and Wavelets*, Gordon and Breach Publ., Luxembourg, 1995.

[93] V. Katz, *A History of Mathematics*, Harper Collins, New York, 1993.

[94] B.M. Kiernan, *The Development of Galois Theory from Lagrange to Artin*, Arch. Hist. Exact Sci. **8** (1971), 40–154.

[95] F. Klein, *Vorlesungen über das Ikosaeder*, Teubner, Leibzig, 1884; republished by Birkhäuser Verlag, Boston, 1993; Engl. transl. *The Icosahedron*, 2nd. ed., Dover, New York, 1956.

[96] F. Klein, *Geometry*, 3rd ed., Springer-Verlag, Berlin, 1925; published in English transl. by Dover Publ., New York, 1939.

[97] M. Kline, *Mathematical Thought from Ancient to Modern Times*, Oxford University Press, New York, 1972.

[98] W. Knorr, *The Ancient Tradition of Geometric Problems*, Dover, New York, 1993.

[99] W. Knorr, *The Method of Indivisibles in Ancient Geometry*, in *Vita Mathematica: Historical Research and Integration with Teaching* (ed. Ronald Calinger), Mathematical Association of America, Washington, D.C., 1996, 67–86.

[100] E. Kummer, *Collected Papers*, vol. 1, André Weil (ed.), Springer-Verlag, New York, 1975.

[101] J.L. Lagrange, *Refléxions sur la Résolution Algébrique des Equations*, in Oeuvres de Lagrange, v. 3, Gauthier-Villars, Paris, 1869.

[102] R. Laubenbacher and D. Pengelley, *A Reevaluation of Sophie Germain's Work on Fermat's Last Theorem*, in preparation.

[103] R. Laubenbacher, D. Pengelley, and M. Siddoway, *Recovering Motivation in Mathematics: Teaching with Original Sources*, Undergraduate Mathematics Education Trends **6**:4 (Sept. 1994).

[104] R. Laubenbacher and D. Pengelley, *Great Problems of Mathematics: A course Based on Original Sources*, Amer. Math. Monthly **99** (1992), 313–317.

[105] R. Laubenbacher and D. Pengelley, *Mathematical Masterpieces: Teaching with Original Sources*, Vita Mathematica: Historical Research and Integration with Teaching (R. Calinger, ed.), Math. Assoc. of Amer., Washington, D.C., 1996, 257–260.

[106] H. Lebesgue, *Development of the Integral Concept*, translated in *Measure and the Integral* (ed. K.O. May), Holden-Day, San Francisco, 1966, 177–194.

[107] A.M. Legendre, *Sur Quelques Objets d'Analyse Indeterminée et Particulièrement sur le Théorème de Fermat*, Second Supplément (Sept. 1825) to *Théorie des Nombres*, Second Edition, 1808.

[108] A.M. Legendre, *Éléments de Géométrie, avec des Notes*, various editions, Paris, Firmin Didot, 1794– .

[109] A.M. Legendre, *Réflexions sur Différentes Manières de Démontrer la Théorie des Parallèles ou le Théorème sur la Somme des Trois Angles du Triangle*, Mémoires de l'Académie (Royale) des Sciences (de l'Institut de France), 2nd series, volume 12 (1833), 367–410, plus sheet of figures.

[110] G.W. Leibniz, *Mathematische Schriften*, vol. V, C.I. Gerhardt (ed.), Georg Olms, Hildesheim, 1962.

[111] *Lexikon Bedeutender Mathematiker*, Verlag Harri Deutsch, Thun / Frankfurt (M.), 1990.

[112] J. Lützen, *Joseph Liouville*, Springer-Verlag, New York, 1990.

[113] M.S. Mahoney, *The Mathematical Career of Pierre de Fermat*, Princeton University Press, Princeton, 1994.

[114] E. Maor, *e, The Story of a Number*, Princeton University Press, Princeton, NJ, 1994.

[115] D.A. Marcus, *Number Fields*, Springer-Verlag, New York, 1977.

[116] J.-C. Martzloff, *A History of Chinese Mathematics*, Springer-Verlag, Berlin, 1997.

[117] B. Mazur, *Number Theory as Gadfly*, Amer. Math. Monthly **98** (1991), 593–610.

[118] K. Menger, *Dimensionstheorie*, Teubner, Leipzig, 1928.

[119] K. Menninger, *Number Words and Number Symbols, A Cultural History of Numbers*, The MIT Press, Cambridge, 1969.

[120] H. Meschkowski, *Probleme des Unendlichen*, Vieweg u. Sohn, Braunschweig, 1967.

[121] H. Meschkowski, *Ways of Thought of Great Mathematicians*, Holden-Day Publ., San Francisco, 1964.

[122] H. Midonick, *The Treasury of Mathematics*, Philosophical Library, Inc., New York, 1965.

[123] A.W. Moore, *The Infinite*, Routledge, New York, 1991.

[124] G.H. Moore, *Zermelo's Axiom of Choice: Its Origins, Development and Influence*, Springer-Verlag, New York, 1982.

[125] G.H. Moore, *Towards a History of Cantor's Continuum Problem*, in The History of Modern Mathematics I, pp. 79–121, J. McCleary and D. Rowe (eds.), Academic Press, San Diego, 1989.

[126] L.J. Mordell, *Three Lectures on Fermat's Last Theorem*, Chelsea Publishing Co., New York, 1962.

[127] O. Neugebauer, *The Exact Sciences in Antiquity*, 2nd edition, Dover, New York, 1969.

[128] O. Neugebauer, *Mathematische Keilschrifttexte*, Quellen und Studien, A 3, Berlin, 1935.

[129] O. Neugebauer and A. Sachs, *Mathematical Cuneiform Texts*, Amer. Oriental Soc., New Haven, Conn., 1945.

[130] J.R. Newman, *The World of Mathematics*, Simon and Schuster, New York, 1956.

[131] O. Ore, *Number Theory and its History*, Dover, New York, 1988.

[132] L.M. Osen, *Women in Mathematics*, The MIT Press, Cambridge, 1974.

[133] H. Poincaré, *Oeuvres*, Gauthier-Villars, Paris, 1916–1956.

[134] Proclus, *A Commentary on the First Book of Euclid's Elements*, tr. G.R. Morrow, Princeton University Press, 1970.

[135] W. Purkert and H.J. Ilgauds, *Georg Cantor, 1845–1918*, Birkhäuser, Basel, 1987.

[136] R. Rashed (ed.), *Diophantus, Lés Arithmétiques. Livres IV–VII, Zweisprachige Ausg. (Oeuvres de Diophante, vol. III et IV)*, Les Belles Lettres, Paris, 1984.

[137] P. Ribenboim, *13 Lectures on Fermat's Last Theorem*, Springer-Verlag, New York, 1979.

[138] K. Ribet and B. Hayes, *Fermat's Last Theorem and Modern Arithmetic*, American Scientist **82** (1994), 144–156.

[139] B. Riemann, *Collected Works*, H. Weber (ed.), Dover, New York, 1953.

[140] A. Robinson, *Non-standard Analysis*, 2nd edition, North-Holland, Amsterdam, 1974.

[141] A. Robinson, *Numbers—What Are They and What Are They Good For*, Yale Scientific Magazine **47**, 14–16.

[142] S. Rockey and M. Paolillo, *Bibliography of Collected Works of Mathematicians*, Cornell University Mathematics Library, Ithaca, New York, 1997; http://math.cornell.edu/ library/.

[143] T. Rothman, *Genius and Biographers—the Fictionalization of Evariste Galois*, Amer. Math. Monthly **89** (1982), 84–106.

[144] B.A. Rosenfeld, *A History of Non-Euclidean Geometry: Evolution of the Concept of Geometric Space*, Springer-Verlag, New York, 1988.

[145] R. Rucker, *Infinity and the Mind*, Bantam Books, New York, 1983.

[146] P. Ruffini, *Opere Matematiche* (3 vols.) E. Bortolotti (ed.), Ed. Cremonese della Casa Editrice Perella, Rome, 1953–1954.

[147] S. Russ, *A Translation of Bolzano's Paper on the Intermediate Value Theorem*, Historia Mathematica 7 (1980), 156–185.

[148] S. Russ, "Bolzano's Analytic Programme," *Mathematical Intelligencer* 14, No. 3 (1992), 45–53.

[149] B. Russell, *Principles of Mathematics*, W. W. Norton & Co., New York, 1903.

[150] G. Saccheri, Euclides Vindicatus, Milano, 1733, reprinted by Chelsea Publ. Co., New York, 1986.

[151] C. Schilling and J. Kramer, *Wilhelm Olbers, Sein Leben und Seine Werke*, vol. I, Berlin, 1900/09.

[152] I. Schneider, *Die Situation der mathematischen Wissenschaften vor und zu Beginn der wissenschaftlichen Laufbahn von Gauß*, in *Carl Friedrich Gauß (1777–1855)*, I. Schneider (ed.), Minerva Publikation, Munich, 1981.

[153] M.R. Schroeder, *Number Theory in Science and Communication*, 2nd ed., Springer-Verlag, New York, 1990.

[154] H.A. Schwarz, *Über diejenigen Fälle, in welchen die Gaußische hypergeometrische Reihe eine algebraische Function ihres vierten elementes darstellt*, J. Reine u. Angewandte Mathematik 75 (1872), 292–335.

[155] J. Sesiano (ed.), *Books IV to VII of Diophantus' Arithmetica in the Arabic Translation Attributed to Qustā Ibn Lūqā*, Springer-Verlag, New York, 1982.

[156] M.-K. Siu, *The ABCD of Using History of Mathematics in the Classroom*, Bulletin of the Hong Kong Mathematical Society 1:1 (1997), 143–154.

[157] G. Simmons, *Calculus Gems: Brief Lives and Memorable Mathematics*, McGraw-Hill, New York, 1992.

[158] S. Singh and K. Ribet, *Fermat's Last Stand*, Scientific American, Nov. 1997.

[159] D.E. Smith, *History of Mathematics*, Dover, New York, 1958.

[160] D.E. Smith, *A Source Book in Mathematics*, Dover, New York, 1959.

[161] D. Solow, *How to Read and Do Proofs* (2nd ed.), J. Wiley & Sons, New York, 1990.

[162] S. Stahl, *The Poincaré Half-Plane; A Gateway to Modern Geometry*, Jones and Bartlett, Boston, 1993.

[163] I.N. Stewart and D.O. Tall, *Algebraic Number Theory*, Chapman & Hall, London, 1987.

[164] J. Stillwell, *Sources of Hyperbolic Geometry*, American Mathematical Society, Providence, Rhode Island, 1996.

[165] P. Straffin, *Liu Hui and the First Golden Age of Chinese Mathematics*, Mathematics Magazine 71 (1998), 163–181.

[166] D.J. Struik, *A Source Book in Mathematics, 1200–1800*, Princeton University Press, Princeton, 1986.

[167] F. Swetz, J. Fauvel, O. Bekken, B. Johansson, V. Katz (eds.), *Learn from the Masters!*, Math. Assoc. of Amer., Washington, D.C., 1995.

[168] F. Swetz (ed.), *From Five Fingers to Infinity: A Journey Through the History of Mathematics*, Open Court, Chicago, 1994.

[169] J.-P. Tignol, *Galois' Theory of Algebraic Equations*, John Wiley & Sons, New York, 1980.

[170] I. Todhunter (ed.), *Euclid's Elements*, J.M. Dent & Sons Ltd., London, 1862.

[171] C. Vanden Eynden, *Elementary Number Theory*, Random House, New York, 1987.

[172] F. Viète, *The Analytic Art*, Kent State University Press, Kent, Ohio, 1983.

[173] B.L. Van Der Waerden, *Science Awakening*, P. Noordhoff Ltd., Groningen, Holland, 1954.

[174] B.L. Van Der Waerden, *Die Galoische Theorie von Heinrich Weber bis Emil Artin*, Arch. Hist. Exact Sci. **9** (1972), 240–248.

[175] B.L. Van Der Waerden, *Modern Algebra*, F. Ungar Publ., New York, 1949.

[176] S. Wagon, *The Banach–Tarski Paradox*, Cambridge University Press, New York, 1993.

[177] A. Weil, *Number Theory: An Approach Through History; From Hammurapi to Legendre*, Birkhäuser, Boston, 1983.

[178] E. Winter (ed.), *Bernard Bolzano: Ausgewählte Schriften*, Union Verlag, Berlin, 1976.

[179] H. Wussing, *The Genesis of the Abstract Group Concept*, MIT Press, Cambridge, 1984.

[180] E. Zermelo, *Untersuchungen über die Grundlagen der Mengenlehre. I.*, Math. Annalen **65** (1908), 261–281.

[181] H.G. Zeuthen, *Geschichte der Mathematik im Altertum und Mittelalter*, Teubner, Leipzig, 1912.

Credits

Front cover: (*background*) courtesy Niedersächsische Staats und Universitätsbibliothek Göttingen, Abteilung für Handschriften und seltene Drücke; (*upper left*) from J. Dauben, *Georg Cantor: His Mathematics and Philosophy of the Infinite*, Princeton University Press, Princeton, NJ, 1990; courtesy of Ivor Grattan-Guinness; (*lower left*) from D.E. Smith, *History of Mathematics*, vol. 1, Dover Publications, New York, 1958; by permission; (*center*) from J. Stillwell, *Mathematics and Its History*, Springer-Verlag, New York, 1989; by permission; (*upper right*) from H. Meschkowski, *Denkweisen Grosser Mathematiker*, Friedrich Vieweg & Sohn, Braunschweig, 1990; by permission; (*lower right*) from L. Rigatelli, *Evariste Galois, 1811–1832*, Birkhäuser Verlag, Basel, 1996; by permission.

5: Photo 1.1 from D.E. Smith, *History of Mathematics*, vol. 1, Dover Publications, New York, 1958; by permission.

12: Photo 1.2 from W. Bühler, *Gauss: A Biographical Study*, Springer-Verlag, New York, 1981; by permission.

16: Photo 1.3 from H. Meschkowski, *Denkweisen Grosser Mathematiker*, Friedrich Vieweg & Sohn, Braunschweig, 1990; courtesy of Konrad Jacobs.

18: Photo 1.4 from H. Meschkowski, *Denkweisen Grosser Mathematiker*, Friedrich Vieweg & Sohn, Braunschweig, 1990; by permission from Deutsches Museum, München.

25: Photo 1.5 from D. Struik, *A Concise History of Mathematics*, Dover Publications, New York, 1967; by permission.

32: Photo 1.6, see credit for front cover, center.

35: Photo 1.7 from H. Meschkowski, *Denkweisen Grosser Mathematiker*, Friedrich Vieweg & Sohn, Braunschweig, 1990; by permission.

43: Photo 1.8 from H. Meschkowski, *Denkweisen Grosser Mathematiker*, Friedrich Vieweg & Sohn, Braunschweig, 1990; by permission from Deutsches Museum, München.

47: Photo 1.9 from *Oeuvres de Henri Poincaré*, vol. XI, Gauthiers-Villars, Paris, 1956; by permission from ESME, 23 rue Linois, 75724 Paris Cedex 15, France.

51: Figure 1.25 from J. Stillwell, *Sources of Hyperbolic Geometry*, American Mathematical Society, Providence, RI, 1996; by permission.

53: Figure 1.27 M.C. Escher's "Circle Limit IV," © 1998 Cordon Art B.V., Baarn, The Netherlands. All rights reserved; by permission.

65: Photo 2.1 from H. Meschkowski, *Problemgeschichte der Mathematik III*, B.I. Wissenschaftsverlag, Zürich, 1986.

68: Photo 2.2 from Kurt Gödel, *Collected Works*, vol. I, Oxford University Press, 1986; courtesy of the Institute for Advanced Study, Princeton, NJ, and the University of Notre Dame.

70: Photo 2.3 from H. Meschkowski, *Denkweisen Grosser Mathematiker*, Friedrich Vieweg & Sohn, Braunschweig, 1990; by permission from Deutsches Museum, München.

75: Photo 2.4, see credit for front cover, upper left.

90: Photo 2.5 from G. Moore, *Zermelo's Axiom of Choice: Its Origins, Development, and Influence*, Springer-Verlag, New York, 1982; by permission.

96: Photo 3.1 from C. Boyer, *A History of Mathematics*, John Wiley & Sons, 1968; by permission from Städelsches Kunstinstitut und Städtische Galerie, Frankfurt am Main.

105: Photo 3.2 from C. Boyer, *A History of Mathematics*, John Wiley & Sons, New York, 1968; by permission.

109: Photo 3.3 from H. Meschkowski, *Denkweisen Grosser Mathematiker*, Friedrich Vieweg & Sohn, Braunschweig,1990; by permission from Deutsches Museum, München.

126: Photo 3.4 and Figure 3.7 from Bonaventura Cavalieri, *Exercitationes Geometricae Sex*, Bologna, 1647, reprinted by Unione Matematica Italiana, Istituto Statale d'Arte di Urbino, 1980.

128: Figure 3.8 from Bonaventura Cavalieri, *Exercitationes Geometricae Sex*, Bologna, 1647, reprinted by Unione Matematica Italiana, Istituto Statale d'Arte di Urbino, 1980.

130: Photo 3.5, see credit for front cover, lower left.

139: Photo 3.6 from D.E. Smith, *History of Mathematics*, vol. 1, Dover Publications, New York, 1958; by permission.

151: Photo 3.7 from J. Dauben, *Abraham Robinson: The Creation of Nonstandard Analysis: A Personal and Mathematical Odyssey*, Princeton University Press, Princeton, NJ, 1995; courtesy of Renée Robinson.

157: Photo 4.1 from K. Barner, "Wolfskehl and the Wolfskehl Prize," *Notices, American Mathematical Society*, vol. 44 (November 1997), pp. 1294–1303; courtesy of Klaus Barner.

158: Photo 4.2, see credit for front cover, upper right.

162: Photo 4.3 from D.E. Smith, *History of Mathematics*, vol. 2, Dover Publications, New York, 1958; by permission.

163: Photo 4.4 from A. Weil, *Number Theory: An Approach Through History from Hammurapi to Legendre*, Birkhäuser, Boston, 1984; by permission.

164: Photo 4.5 from A. Weil, *Number Theory: An Approach Through History from Hammurapi to Legendre*, Birkhäuser, Boston, 1984; by permission.

171: Photo 4.6 from Gina Kolata, "Andrew Wiles: A Math Whiz Battles 350-Year-Old Puzzle," *Math Horizons*, Winter 1993, pp. 8–11; by permission from Princeton University Information Services.

173: Photo 4.7 from A. Weil, *Number Theory: An Approach Through History from Hammurapi to Legendre*, Birkhäuser, Boston, 1984; by permission from the American Oriental Society.

174: Photo 4.8 from H. Meschkowski, *Denkweisen Grosser Mathematiker*, Friedrich Vieweg & Sohn, Braunschweig, 1990; by permission from Deutsches Museum, München.

175: Photo 4.9 from D.E. Smith, *History of Mathematics*, vol. 1, Dover Publications, New York, 1958; by permission.

180: Photo 4.10 from H. Meschkowski, *Denkweisen Grosser Mathematiker*, Friedrich Vieweg & Sohn, Braunschweig, 1990; by permission.

186: Photo 4.11 from L. Bucciarelli, *Sophie Germain: An Essay in the History of the Theory of Elasticity*, D. Reidel, Dordrecht, Holland, 1980; by permission from Lawrence Bucciarelli.

188: Photo 4.12, see credit for front cover, background.

190: Photo 4.13 from MS. FR 9114, p. 92, Bibliothèque Nationale, Paris; by permission.

195: Photo 4.14 from E. Kummer, *Collected Papers*, vol. 2, Springer-Verlag, Berlin, 1975; by permission.

205: Photo 5.1 from J. Fauvel and J. Gray, *The History of Mathematics: A Reader*, The Open University, 1987; by permission from the American Oriental Society.

213: Photo 5.2 from *Acta Mathematica*, vol. 1 (1882).

215: Photo 5.3 from K. Biermann, *Carl F. Gauss: "Der Fürst der Mathematiker" in Briefen und Gesprächen*, Verlag C.H. Beck, München, 1990; by permission from Universitätsbibliothek Leipzig.

218: Photo 5.4 from A. Dick, *E. Noether*, Birkhäuser, Boston, 1981; by permission.

220: Photo 5.5 from C. Boyer, *A History of Mathematics*, John Wiley & Sons, New York, 1968; by permission.

225: Photo 5.6 from T.R. Witmer, *The Great Art, or the Rules of Algebra*, M.I.T. Press, Cambridge, MA, 1968; by permission.

226: Photo 5.7 from H. Meschkowski, *Denkweisen Grosser Mathematiker*, Friedrich Vieweg & Sohn, Braunschweig, 1990; by permission from Deutsches Museum, München.

228: Photo 5.8 from D. Struik, *Sourcebook in Mathematics, 1200–1800*, Princeton University Press, Princeton, NJ, 1986.

234: Photo 5.9 from *Lexikon Bedeutender Mathematiker*, Verlag Harri Deutsch, Frankfurt (M.), 1990.

248: Photo 5.10, see credit for front cover, lower right.

250: Photo 5.11 from H. Meschkowski, *Problemgeschichte der Mathematik III*, B.I. Wissenschaftsverlag, Zürich, 1986.

Index

Anglin: Mathematics: A Concise History and Philosophy.
Readings in Mathematics.

Anglin/Lambek: The Heritage of Thales.
Readings in Mathematics.

Apostol: Introduction to Analytic Number Theory. Second edition.

Armstrong: Basic Topology.

Armstrong: Groups and Symmetry.

Axler: Linear Algebra Done Right. Second edition.

Beardon: Limits: A New Approach to Real Analysis.

Bak/Newman: Complex Analysis. Second edition.

Banchoff/Wermer: Linear Algebra Through Geometry. Second edition.

Berberian: A First Course in Real Analysis.

Bix: Conics and Cubics: A Concrete Introduction to Algebraic Curves.

Brémaud: An Introduction to Probabilistic Modeling.

Bressoud: Factorization and Primality Testing.

Bressoud: Second Year Calculus.
Readings in Mathematics.

Brickman: Mathematical Introduction to Linear Programming and Game Theory.

Browder: Mathematical Analysis: An Introduction.

Buskes/van Rooij: Topological Spaces: From Distance to Neighborhood.

Callahan: The Geometry of Spacetime: An Introduction to Special and General Relativity.

Carter/van Brunt: The Lebesgue-Stieltjes: A Practical Introduction

Cederberg: A Course in Modern Geometries.

Childs: A Concrete Introduction to Higher Algebra. Second edition.

Chung: Elementary Probability Theory with Stochastic Processes. Third edition.

Cox/Little/O'Shea: Ideals, Varieties, and Algorithms. Second edition.

Croom: Basic Concepts of Algebraic Topology.

Curtis: Linear Algebra: An Introductory Approach. Fourth edition.

Devlin: The Joy of Sets: Fundamentals of Contemporary Set Theory. Second edition.

Dixmier: General Topology.

Driver: Why Math?

Ebbinghaus/Flum/Thomas: Mathematical Logic. Second edition.

Edgar: Measure, Topology, and Fractal Geometry.

Elaydi: An Introduction to Difference Equations. Second edition.

Exner: An Accompaniment to Higher Mathematics.

Exner: Inside Calculus.

Fine/Rosenberger: The Fundamental Theory of Algebra.

Fischer: Intermediate Real Analysis.

Flanigan/Kazdan: Calculus Two: Linear and Nonlinear Functions. Second edition.

Fleming: Functions of Several Variables. Second edition.

Foulds: Combinatorial Optimization for Undergraduates.

Foulds: Optimization Techniques: An Introduction.

Franklin: Methods of Mathematical Economics.

Frazier: An Introduction to Wavelets Through Linear Algebra.

Gordon: Discrete Probability.

Hairer/Wanner: Analysis by Its History.
Readings in Mathematics.

Halmos: Finite-Dimensional Vector Spaces. Second edition.

Halmos: Naive Set Theory.

Hämmerlin/Hoffmann: Numerical Mathematics.
Readings in Mathematics.

Harris/Hirst/Mossinghoff: Combinatorics and Graph Theory.

Hartshorne: Geometry: Euclid and Beyond.

Hijab: Introduction to Calculus and Classical Analysis.

Hilton/Holton/Pedersen: Mathematical Reflections: In a Room with Many Mirrors.

Iooss/Joseph: Elementary Stability and Bifurcation Theory. Second edition.

Isaac: The Pleasures of Probability. *Readings in Mathematics.*

James: Topological and Uniform Spaces.

Jänich: Linear Algebra.

Jänich: Topology.

Kemeny/Snell: Finite Markov Chains.

Kinsey: Topology of Surfaces.

Klambauer: Aspects of Calculus.

Lang: A First Course in Calculus. Fifth edition.

Lang: Calculus of Several Variables. Third edition.

Lang: Introduction to Linear Algebra. Second edition.

Lang: Linear Algebra. Third edition.

Lang: Undergraduate Algebra. Second edition.

Lang: Undergraduate Analysis.

Lax/Burstein/Lax: Calculus with Applications and Computing. Volume 1.

LeCuyer: College Mathematics with APL.

Lidl/Pilz: Applied Abstract Algebra. Second edition.

Logan: Applied Partial Differential Equations.

Macki-Strauss: Introduction to Optimal Control Theory.

Malitz: Introduction to Mathematical Logic.

Marsden/Weinstein: Calculus I, II, III. Second edition.

Martin: The Foundations of Geometry and the Non-Euclidean Plane.

Martin: Geometric Constructions.

Martin: Transformation Geometry: An Introduction to Symmetry.

Millman/Parker: Geometry: A Metric· Approach with Models. Second edition.

Moschovakis: Notes on Set Theory.

Owen: A First Course in the Mathematical Foundations of Thermodynamics.

Palka: An Introduction to Complex Function Theory.

Pedrick: A First Course in Analysis.

Peressini/Sullivan/Uhl: The Mathematics of Nonlinear Programming.

Prenowitz/Jantosciak: Join Geometries.

Priestley: Calculus: A Liberal Art. Second edition.

Protter/Morrey: A First Course in Real Analysis. Second edition.

Protter/Morrey: Intermediate Calculus. Second edition.

Roman: An Introduction to Coding and Information Theory.

Ross: Elementary Analysis: The Theory of Calculus.

Samuel: Projective Geometry. *Readings in Mathematics.*

Scharlau/Opolka: From Fermat to Minkowski.

Schiff: The Laplace Transform: Theory and Applications.

Sethuraman: Rings, Fields, and Vector Spaces: An Approach to Geometric Constructability.

Sigler: Algebra.

Silverman/Tate: Rational Points on Elliptic Curves.

Simmonds: A Brief on Tensor Analysis. Second edition.

Singer: Geometry: Plane and Fancy.

Singer/Thorpe: Lecture Notes on Elementary Topology and Geometry.

Smith: Linear Algebra. Third edition.

Smith: Primer of Modern Analysis. Second edition.

Stanton/White: Constructive Combinatorics.

Stillwell: Elements of Algebra: Geometry, Numbers, Equations.

Stillwell: Mathematics and Its History.

Stillwell: Numbers and Geometry. *Readings in Mathematics.*

Strayer: Linear Programming and Its Applications.

(continued on next page)

Undergraduate Texts in Mathematics

Anglin: Mathematics: A Concise History and Philosophy.
Readings in Mathematics.

Anglin/Lambek: The Heritage of Thales.
Readings in Mathematics.

Apostol: Introduction to Analytic Number Theory. Second edition.

Armstrong: Basic Topology.

Armstrong: Groups and Symmetry.

Axler: Linear Algebra Done Right. Second edition.

Beardon: Limits: A New Approach to Real Analysis.

Bak/Newman: Complex Analysis. Second edition.

Banchoff/Wermer: Linear Algebra Through Geometry. Second edition.

Berberian: A First Course in Real Analysis.

Bix: Conics and Cubics: A Concrete Introduction to Algebraic Curves.

Brémaud: An Introduction to Probabilistic Modeling.

Bressoud: Factorization and Primality Testing.

Bressoud: Second Year Calculus.
Readings in Mathematics.

Brickman: Mathematical Introduction to Linear Programming and Game Theory.

Browder: Mathematical Analysis: An Introduction.

Buskes/van Rooij: Topological Spaces: From Distance to Neighborhood.

Callahan: The Geometry of Spacetime: An Introduction to Special and General Relativity.

Carter/van Brunt: The Lebesgue-Stieltjes: A Practical Introduction

Cederberg: A Course in Modern Geometries.

Childs: A Concrete Introduction to Higher Algebra. Second edition.

Chung: Elementary Probability Theory with Stochastic Processes. Third edition.

Cox/Little/O'Shea: Ideals, Varieties, and Algorithms. Second edition.

Croom: Basic Concepts of Algebraic Topology.

Curtis: Linear Algebra: An Introductory Approach. Fourth edition.

Devlin: The Joy of Sets: Fundamentals of Contemporary Set Theory. Second edition.

Dixmier: General Topology.

Driver: Why Math?

Ebbinghaus/Flum/Thomas: Mathematical Logic. Second edition.

Edgar: Measure, Topology, and Fractal Geometry.

Elaydi: An Introduction to Difference Equations. Second edition.

Exner: An Accompaniment to Higher Mathematics.

Exner: Inside Calculus.

Fine/Rosenberger: The Fundamental Theory of Algebra.

Fischer: Intermediate Real Analysis.

Flanigan/Kazdan: Calculus Two: Linear and Nonlinear Functions. Second edition.

Fleming: Functions of Several Variables. Second edition.

Foulds: Combinatorial Optimization for Undergraduates.

Foulds: Optimization Techniques: An Introduction.

Franklin: Methods of Mathematical Economics.

Frazier: An Introduction to Wavelets Through Linear Algebra.

Gordon: Discrete Probability.

Hairer/Wanner: Analysis by Its History.
Readings in Mathematics.

Halmos: Finite-Dimensional Vector Spaces. Second edition.

Halmos: Naive Set Theory.

Hämmerlin/Hoffmann: Numerical Mathematics.
Readings in Mathematics.

Harris/Hirst/Mossinghoff: Combinatorics and Graph Theory.

Hartshorne: Geometry: Euclid and Beyond.

Hijab: Introduction to Calculus and Classical Analysis.

Printed in the United States
by Baker & Taylor Publisher Services